PARTIAL DIFFERENTIAL EQUATIONS

This book is an attempt to make available to the student a coherent modern view of the theory of partial differential equations. Here equations of the first order and linear second order equations are treated with the tensor calculus, which combines generality and insight, in mind. Since the book is self-contained, much of the material is classical, but an effort has been made to achieve a modern outlook on these topics. A number of significant recent developments are introduced, and treated in relation to the natural background formed by geometry and physics.

Special features of the exposition are: *(a)* the simplified general treatment of first order equations; *(b)* the geometrical foundations of the theory of linear second order equations; *(c)* unified treatment of boundary value problems and related topics by integral equations; *(d)* the theory of generalized hyperbolic potentials.

G. F. D. DUFF was born and educated in Toronto, taking his M.A. at the University of Toronto in 1949. He obtained his Ph.D. from Princeton, and after a year's teaching at the Massachusetts Institute of Technology he returned to the University of Toronto where he is now Professor in the Department of Mathematics. His mathematical interests are centred in the theory of differential equations, and he has also worked in the theory of harmonic integrals.

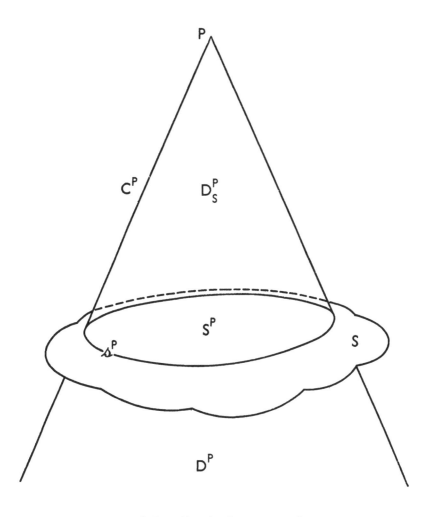

Cauchy's problem for the wave equation

MATHEMATICAL EXPOSITIONS

Volumes Published

MATHEMATICAL EXPOSITIONS No. 9

PARTIAL DIFFERENTIAL EQUATIONS

BY

G. F. D. DUFF

PROFESSOR OF MATHEMATICS
UNIVERSITY OF TORONTO

UNIVERSITY OF TORONTO PRESS

TORONTO

PREFACE

THIS book is intended as an introduction to the theory of partial differential equations of the first and second orders, which will be useful to the prospective student of modern developments, and also to those who desire a detailed background for the traditional applications. The discussion is restricted to equations of the first order and linear equations of the second order, with one dependent variable, so that a large part of the work is severely classical. However, some of the material is relatively modern, though the methods and concepts used have been restricted to those of classical analysis. The notion of invariance under transformations of the independent variables is emphasized, and the exposition is self-contained as regards tensor calculus. The book might therefore be read by a student who has a good background in ordinary differential equations.

On the one hand, this treatment is intended to serve as preparation for such topics as general theories of the integration of differential systems, harmonic integrals, or the study of differential operators in function spaces. On the other hand, the equations treated here are useful in many branches of applied mathematics, and might well be studied from a more general standpoint by those who are concerned with the applications. As a branch of mathematical knowledge, our subject is notable for its many contacts with other branches of pure and applied mathematics. This interaction with its environment has produced not just a mass of diverse particular facts, but a well-organized and tightly knit theory. The aim is to present this aspect of the subject in a reasonably accessible form.

Each chapter begins with a summary wherein the motivation and order of topics within the chapter are described. Exercises are included in nearly every section; many of them are particular applications, though a few suggest further developments. A short bibliography, mainly of treatises and monographs, is appended, together with some references from each of the chapters. For the sake of brevity I have not attempted to state precise conditions of regularity for many of the results which are discussed herein, feeling that the student who wishes to pursue such questions will in any case turn to the standard treatises.

My colleagues Professors J. D. Burk and A. Robinson read the manuscript at various stages of completion and contributed many helpful criticisms, for which I make grateful acknowledgement. It is also a pleasure to thank Professor D. C. Spencer of Princeton University for several valuable comments.

G. F. D. DUFF

Toronto, March 1955

CONTENTS

I

DIFFERENTIAL EQUATIONS AND THEIR SOLUTIONS

1.1. Some definitions and examples. A relation of the form

$$F\left(x^1, x^2, ..., x^N, u, \frac{\partial u}{\partial x^1}, ..., \frac{\partial u}{\partial x^N}, \frac{\partial^2 u}{(\partial x^1)^2}, ..., \frac{\partial^k u}{(\partial x^N)^k}\right) = 0 \quad (1.1.1)$$

is known as a partial differential equation. In this equation the quantities $x^1, x^2, ..., x^N$ are independent variables; they range over certain sets of real values; or, as is often said for convenience, over a domain or region in the 'space' of the independent variables. A single set of numerical values $x^1, ..., x^N$ is usually called a 'point' in that space. The quantity u which appears in the equation plays the role of a dependent variable. That is, u is assumed to be a function $u = f(x^1, x^2, ..., x^N)$ of the independent variables, such that all derivatives of u which appear in (1.1.1) exist.

If we select a function $f(x^1, x^2, ..., x^N)$ which possesses the requisite derivatives but is otherwise arbitrary, and if we form the expression

$$F\left(x^1, x^2, ..., x^N, f(x^1, ..., x^N), \frac{\partial f(x^1, ..., x^N)}{\partial x^1}, ..., \frac{\partial^k f}{(\partial x^N)^k}\right), \quad (1.1.2)$$

we obtain a compounded function which is a function of $x^1, x^2, ..., x^N$. In general, this function will not be identically zero in the domain of the space of the $x^1, ..., x^N$ which we wish to consider. However, if it does happen that the expression (1.1.2) vanishes identically in this domain, then we say that the function is a solution of the partial differential equation (1.1.1). That is, the relation

$$u = f(x^1, ..., x^N), \quad (1.1.3)$$

which is known as an integral relation or integral has as a consequence the truth of the differential relation (1.1.1) among the $x^1, ..., x^N$, u, and the partial derivatives of u with respect to the $x^1, ..., x^N$.

Throughout this book we shall limit the discussion to equations which, like (1.1.1), contain a single dependent variable u. The number N of independent variables will always be two or more, since if there were but one independent variable, the equation (1.1.1) would be an ordinary differential equation. We shall assume that the reader is familiar with certain properties of ordinary differential equations which, from time to time, will be needed in the exposition. The theory of partial differential equations is a natural extension of the theory of ordinary differential equations which in turn grew from the differential and integral calculus. The development of all these subjects has been powerfully stimulated by physical problems and applications which require analytical methods. Among these the questions which lead to partial differential equations are second to none in variety and usefulness. Our present purpose being, however, to study partial differential equations for their own sake, we shall employ an approach and motivation which is essentially mathematical.

We recall certain basic definitions. The order of a differential equation is the order of the highest derivative which appears in the equation. There are also definitions which refer to the algebraic structure of the equation. If the function F in (1.1.1) is linear in u and all of the derivatives of u which appear, the equation is said to be linear. Again, the linear equation is homogeneous if no term independent of u is present.

Linear homogeneous equations have a special property, which greatly facilitates their treatment, namely, that a constant multiple of a solution, or the sum of two or more solutions, is again a solution. Thus known solutions can be superposed, to build up new solutions. Not only sums, but integrals over solutions containing a parameter may be used, and for this reason it is often possible to find explicit formulae for the solutions of problems which involve linear equations. In comparison, the treatment of non-linear equations is much more difficult, and the available results less comprehensive.

Frequently it is possible to simplify the form of a differential equation by a suitable transformation of the dependent or independent variables. Certain standard forms have therefore been adopted and closely studied, since their properties carry over to equations of an apparently more general character. A well-known example of an equation in two independent variables (denoted by x and y for convenience) which may be written in either of two different forms is

$$\frac{\partial^2 u}{\partial x^2} - \frac{\partial^2 u}{\partial y^2} = f(x, y). \tag{1.1.4}$$

If in this equation we set $x+y = \xi$, $x-y = \eta$, then the rules for calculation with partial derivatives show that, as a function of ξ, η, the dependent variable u satisfies

$$\frac{\partial^2 u}{\partial \xi \partial \eta} = \tfrac{1}{4} f\left(\frac{\xi+\eta}{2}, \frac{\xi-\eta}{2}\right). \tag{1.1.5}$$

When considering a differential equation such as (1.1.1), we must guard against the presumption that it has any solutions at all, until a proof of this has been supplied. In this connexion it might be necessary to specify the class of functions from which solutions will be selected. For example, the equation

$$\left(\frac{\partial u}{\partial x}\right)^2 + \left(\frac{\partial u}{\partial y}\right)^2 + 1 = 0$$

clearly has no real-valued solutions whatever; but there are infinitely many complex-valued solutions of the form

$$u = \alpha x + \beta y + \gamma,$$

where α, β, γ are any complex constants such that $\alpha^2 + \beta^2 + 1 = 0$. For simplicity, we shall hereafter assume that all equations, functions, and solutions are real-valued, unless the contrary is explicitly indicated.

Let us recall that an ordinary differential equation usually possesses infinitely many distinct solution functions. When the equation can be integrated in explicit analytical form, there appear constants of integration to which may be assigned numerical values in infinitely many ways. For instance, the equation of the second order $u'' = f(x)$ has as solution

$$u(x) = \int\limits_0^x (x-t)f(t)\,dt + Ax + B,$$

where A and B are arbitrary constants. Thus the choice of a single solution function is made, in this case, by choosing two numbers. However, the solution of the most general form of a partial differential equation will usually contain arbitrary functions. Indeed, in the above example, if u depended on an additional variable y, the quantities A and B could be chosen as any functions of y.

Examples of the functional arbitrariness of general forms of solutions of partial differential equations are easily found. For example,

$$u = f(y - x^2),$$

where $f(t)$ is any once differentiable function, is a solution of the equation of the first order

$$\frac{\partial u}{\partial x} + 2x\frac{\partial u}{\partial y} = 0.$$

A general solution of the second-order equation (1.1.5) is

$$u(\xi, \eta) = \tfrac{1}{4} \int\limits_0^\xi \int\limits_0^\eta f\left(\frac{s+t}{2}, \frac{s-t}{2}\right) ds\,dt + A(\xi) + B(\eta), \qquad (1.1.6)$$

in which there appear two arbitrary functions.

Conversely, it is often possible to derive a partial differential equation from a relation involving arbitrary functions. This can be done in the two cases above. As a further example consider the general equation

$$u = f(x^2+y^2)$$

of the surfaces of revolution about the u-axis in the Euclidean space of coordinates x, y, u. Suppose that f is differentiable; then we see easily that

$$y \frac{\partial u}{\partial x} - x \frac{\partial u}{\partial y} = 0.$$

This partial differential equation of the first order characterizes the surfaces of revolution by a geometric property, namely, that the normal line to the surface meets the u-axis. Conversely, any integral of the partial differential equation has the form given above. The reader can easily verify this by transforming the equation to polar coordinates r, θ in the x, y plane.

Another equation of the first order which leads to a functional relation as integral is

$$\frac{\partial u}{\partial x} \frac{\partial g(x,y,u)}{\partial y} - \frac{\partial u}{\partial y} \frac{\partial g(x,y,u)}{\partial x} = 0, \tag{1.1.7}$$

where $g(x,y,u)$ is a known function of the three variables x, y, and u. Let

$$f(x,y) = g\{x, y, u(x,y)\},$$

then the Jacobian

$$\begin{aligned}
\frac{\partial(u,f)}{\partial(x,y)} &= \frac{\partial u}{\partial x} \frac{\partial f}{\partial y} - \frac{\partial u}{\partial y} \frac{\partial f}{\partial x} \\
&= \frac{\partial u}{\partial x}\left(\frac{\partial g}{\partial y} + \frac{\partial g}{\partial u}\frac{\partial u}{\partial y}\right) - \frac{\partial u}{\partial y}\left(\frac{\partial g}{\partial x} + \frac{\partial g}{\partial u}\frac{\partial u}{\partial x}\right) \\
&= \frac{\partial u}{\partial x} \frac{\partial g}{\partial y} - \frac{\partial u}{\partial y} \frac{\partial g}{\partial x}
\end{aligned}$$

vanishes in view of the differential equation. Therefore there must be some functional relation independent of x and y which subsists between the two functions u and f of these two independent variables. Consequently any integral of the partial differential equation is defined implicitly by a relation

$$u = F\{g(x,y,u)\}.$$

The reader will easily verify that any such function u does satisfy the differential equation. Note that this equation is not linear.

Returning to equations of the second order, let us consider the homogeneous equation corresponding to (1.1.4), namely

$$\frac{\partial^2 u}{\partial x^2} - \frac{\partial^2 u}{\partial y^2} = 0. \tag{1.1.8}$$

This is the wave equation which governs the vibrations of a stretched string. From (1.1.6) and the transformation which leads from (1.1.4) to (1.1.5) we see that

$$u = f(x+y)+g(x-y) \tag{1.1.9}$$

is the most general solution of the equation. If, say, y is interpreted as a time variable, as is appropriate for the string, the two terms of this solution represent waves of arbitrary form but fixed velocity, one wave travelling in each direction along the string.

If we admit complex numbers and replace y by iy ($i^2 = -1$) in (1.1.8) we see that the equation becomes

$$\frac{\partial^2 u}{\partial x^2} + \frac{\partial^2 u}{\partial y^2} = 0, \tag{1.1.10}$$

and has the solution $\quad u = f(x+iy)+g(x-iy). \tag{1.1.11}$

This form of solution suggests the introduction of a complex variable $z = x+iy$, and of its complex conjugate $\bar{z} = x-iy$. It may be remembered that the real and imaginary parts of an analytic function

$$f(z) = f(x+iy) = u+iv$$

separately satisfy (1.1.10). This is Laplace's equation in the plane, and, partly because of the connexion with complex variable theory just suggested, it has been studied in greater detail than any other partial differential equation.

Exercise 1. Eliminate the arbitrary functions from

(a) $\quad u = x^k f(y/x)$,

(b) $\quad u = f(x^2+y^2+u^2)$,

(c) $\quad u = f(x+\alpha y)+g(x+\beta y)$,

(d) $\quad u = f(x+\alpha y)+xg(x+\alpha y)$,

(e) $\quad u = f(x\cos\alpha + y\sin\alpha + z) + g(x\cos\alpha + y\sin\alpha - z)$.

Exercise 2. Find a solution, containing three disposable functions, of

$$\frac{\partial^3 u}{\partial x(\partial y)^2} - \frac{\partial^3 u}{\partial x^3} = 0.$$

Exercise 3. Find a general form of solution for the equation in three variables

$$\frac{\partial^3 u}{\partial x \partial y \partial z} = f(x, y, z).$$

Exercise 4. The envelope of a one-parameter family of planes in Euclidean space of the variables x, y, u, satisfies

$$\frac{\partial^2 u}{\partial x^2}\frac{\partial^2 u}{\partial y^2} - \left(\frac{\partial^2 u}{\partial x \partial y}\right)^2 = 0.$$

Hint: Write the family of planes

$$u = \lambda x + f(\lambda)y + g(\lambda),$$

differentiate with respect to λ, and show that

$$\frac{\partial u}{\partial x} = f\left(\frac{\partial u}{\partial y}\right).$$

Exercise 5. Separation of variables. The heat equation

$$\frac{\partial u}{\partial t} = \frac{\partial^2 u}{\partial x^2}$$

has solutions of the form $u = \varphi(t)\psi(x)$ where

$$\varphi(t) = \exp\{-\lambda^2(t-\tau)\},$$
$$\psi(x) = \sin\lambda(x-\xi),\ \cos\lambda(x-\xi),$$

where λ, τ, ξ are any constants. Hence, by superposition, the integral

$$u = \int_{-\infty}^{\infty} f(\lambda)e^{-\lambda^2 t}\cos(\lambda x)\,d\lambda$$

is a solution. When $f(\lambda) = 1$, $u = \left(\frac{\pi}{t}\right)^{\frac{1}{2}}\exp\left(-\frac{x^2}{4t}\right)$, $t > 0$.

Exercise 6. Separation of variables. The equation of the first order

$$f_1(x)\left(\frac{\partial u}{\partial x}\right)^2 + f_2(y)\left(\frac{\partial u}{\partial y}\right)^2 = g_1(x) + g_2(y)$$

has solutions of the form $u = \varphi(x) + \psi(y)$. Find a family of these solutions which depends upon two arbitrary constants.

Exercise 7. Show that the equation $u_t + uu_x = u_{xx}$ can be transformed to the linear form $z_t = z_{xx} + f(t)z$ by introducing a new dependent variable z defined by the relation $u = -2(\log z)_x$.

1.2. The classification of equations and their solutions. The earliest and most obvious classification of partial differential equations was made on a formal basis. Such properties as the number of independent or dependent variables, the order of the equation, and its algebraic or functional structure, are necessarily the first to be considered. Further classifications have been made according to the methods which are available to treat special types of partial differential equations. For instance, linear equations with constant coefficients may be said to form a category of equations of this sort, since there are methods for finding explicit solutions of these equations, which methods fail when applied to equations with variable coefficients.

However, there is still another important principle by which partial differential equations may be classified. In all studies of this subject, it is really the properties of solutions of the equations which are important, rather than formal properties of the equation itself. In many cases the solutions of equations formally quite similar have radically different properties, and the equations should therefore be treated as distinct types. Classifications based on properties of the solutions are usually less obvious and also deeper than those based on structural properties of the equations. Probably the most important example of such a classification is the division into types of linear equations of the second order considered in Chapter IV.

It is desirable that a classification should take some account of the manner in which equations can be transformed by a change of independent or dependent variables. If the properties of solutions determine the classification, such a transformation will not alter the type of the partial differential equation.

In most practical applications, as well as in many theoretical problems, it is required to find one particular solution which also satisfies certain definite additional conditions. These auxiliary conditions may take the form of 'initial' or 'boundary' conditions, or both. It happens that certain types of auxiliary conditions are appropriate only to certain corresponding types of partial differential equations, in the sense that only for these equations can a solution satisfying the boundary conditions be proved to exist.

The problems of mathematical physics which lead to partial differential equations usually suggest appropriate auxiliary conditions at the same time. For instance, the problems of potential theory (Newtonian gravitation) lead to elliptic equations and to Dirichlet's problem in which boundary conditions are specified on a closed curve or surface; and this type of problem is appropriate for elliptic equations. Similarly, many problems of wave motion lead to hyperbolic equations and to auxiliary initial conditions which are meaningful for hyperbolic equations. When auxiliary conditions are specified in such a way that there exists one and only one solution of the partial differential equation which satisfies the conditions, then the problem of finding the solution is said to be *correctly set*. It is natural that practical problems should be a guide in suggesting how auxiliary conditions may be found for a given partial differential equation. In many cases when this guide is not available, correct auxiliary conditions are not known. However, such considerations as these, though they often suggest the result to be established, do not provide rigorous proofs that the solution of the problem exists.

In connexion with auxiliary conditions it is necessary to consider not only how many functions are to be specified, and how they are to determine the solution, but also such properties of these functions as continuity or analyticity. The data of physical problems are never exact; we must determine the effect upon the solution of small variations or uncertainties in the specified functions which appear in the auxiliary conditions. If small but arbitrary changes in the data lead to equally small perturbations of the solution, the problem is stable, otherwise it is unstable.

The analytical nature of the solution desired is an important part of any problem in partial differential equations. Both the intrinsic nature of the solution, its properties of continuity and differentiability, and the form in which it is to be expressed may determine what method is most appropriate for the problem at hand. Often the local behaviour of the solution can be deduced from the form of the differential equation, or from known existence theorems. Such knowledge can also influence the form in which the solution functions may be expressed—whether by known functions, series, integrals, implicit functions, or other means. If, for example, a solution is known to possess discontinuities, it is not possible to represent these by using a power series expansion.

Among the functions commonly used to represent mathematical or physical phenomena, those which are analytic stand out as a special class, because of their 'rigid' nature. By an analytic function we mean a real-valued function of real variables, expressible by means of convergent power series expansions in those variables. The 'rigidity' of analytic functions springs from the fact that if such a function, together with all its derivatives, is known at one point, then its values at all other points are fixed by the process of continuation with power series. If a change in value, however small, is made at one point, the values of the analytic function at all other points will, in general, be affected. Thus analytic functions are not very suitable for the representation of physical phenomena, in which events at different points may be quite independent of each other. Most of the partial differential equations which appear in practice are themselves analytic, that is, all functions and coefficients in them are analytic in their various arguments; but by no means are all solutions of all these equations also analytic.

Another feature of the study of solutions of partial differential equations deserves our attention, namely, the division of the theory into its local and global aspects. A differential equation is a statement regarding the value of a function and its derivatives at a point—that is, of the function in an arbitrarily small neighbourhood of the point—and many properties of solutions which can be deduced directly from the equation are local properties such as smoothness and regularity. In the next section will be presented existence theorems, applicable to very general analytic equations, which show that these equations do possess solutions defined in small but finite regions—the neighbourhood of a point or of a surface, for example. These results must also be regarded as local.

To solve most practical questions, however, it is necessary to find solutions defined in a given region of finite, or perhaps even infinite, extent. This is the global or 'in the large' aspect. One approach to it would be to piece together solutions which are defined locally; but this is often not feasible, so that quite different methods are needed. In a certain sense, a problem 'in the large' requires us to have at hand all solutions of the partial differential equation, and to select, by some means, the appropriate one. For this reason we may wish to characterize, or if possible to construct, the whole 'manifold of solutions' of a given equation. With ordinary differential equations, explicit integration is usually the difficult process, while the fitting of boundary conditions is easier. The reverse is often true for partial differential equations—it is not hard to find a general form of solution, but it is difficult to specialize it in order to satisfy auxiliary conditions.

These various concepts may be illustrated by a comparison of the two second-order equations mentioned in § 1, namely the wave equation, and Laplace's equation in two dimensions. The wave equation in the independent variables x, t is

$$\frac{\partial^2 u}{\partial x^2} = \frac{\partial^2 u}{\partial t^2},$$

(1.2.1)

and it has the general solution

$$u = f(x+t) + g(x-t).$$

(1.2.2)

The functions f and g should be assumed twice differentiable, in order that the second derivatives which appear in the equation (1.2.1) should exist.

Higher derivatives of f and g may be permitted to have discontinuities or need not exist. If, for example, $f''(x)$ has a discontinuity at $x = 0$, this discontinuity in the second derivative of the solution will be in evidence on the line $x+t = 0$. Similar discontinuities in $g(x)$ at $x = 0$ are 'propagated' along the line $x-t = 0$. These two lines are called the *characteristics* through the point $x = 0$, $t = 0$.

If a stretched string has, at time $t = 0$, a displacement $u = \varphi(x)$, and a transverse velocity $u_t = \psi(x)$, its motion can be determined by (1.2.1). At $t = 0$, we have, from (1.2.2),

$$u = \varphi(x) = f(x) + g(x),$$

and

$$\frac{\partial u}{\partial t} = \psi(x) = f'(x) - g'(x),$$

so that

$$\int^{x} \psi(\alpha)\, d\alpha = f(x) - g(x).$$

Solving for $f(x)$ and $g(x)$, and substituting the resulting functions in (1.2.2), we see that

$$u(x, t) = \tfrac{1}{2}\left\{ \varphi(x+t) + \varphi(x-t) + \int_{x-t}^{x+t} \psi(\alpha)\, d\alpha \right\}. \tag{1.2.3}$$

This is the solution of the initial value problem for equation (1.2.1). We see that the value of u at the point (x, t) depends only on the values of $u(x, 0) = \varphi(x)$ at the points $(x+t, 0)$ and $(x-t, 0)$; and on the values of $u_t(x, 0) = \psi(x)$ on the interval on the line $t = 0$ between these two points. These points therefore limit at either end the 'domain of dependence' of the value of the solution at the point (x, t). Note that the two characteristics through (x, t) pass one through $(x+t, 0)$, the other through $(x-t, 0)$. The solution formula (1.2.3) shows that the solution is stable in the sense that small variations of φ and ψ result in equally small variations in value of $u(x, t)$ for t positive.

A very different behaviour characterizes the solutions of Laplace's equation

$$\frac{\partial^2 u}{\partial x^2} + \frac{\partial^2 u}{\partial y^2} = 0 \tag{1.2.4}$$

which, as we saw, has the complex-valued general solution

$$u = f(x+iy) + g(x-iy). \tag{1.2.5}$$

The real and imaginary parts of this solution separately satisfy (1.2.4). Solutions of (1.2.4) are known as harmonic functions. The first term $f(x+iy)$ in (1.2.5) has the form of a function $f(z)$ of a complex variable

$z = x+iy$. Now it is known from the theory of functions of a complex variable that if $f(z)$ is differentiable once in the complex sense, it is differentiable any number of times and is in fact an analytic function of the complex variable. Furthermore its real and imaginary parts are real analytic functions of x and y. This strongly suggests, therefore, that all real-valued solutions of (1.2.4) are real analytic functions; and this is in fact true. Thus the two equations we have examined, though similar in form, possess contrasting 'local' properties.

The stability properties of the two equations are also very different. The functions

$$u_n = \frac{1}{n}\sin(nx)\cosh(ny)$$

are harmonic; and on the line $y = 0$ we have $|u_n| \leqslant 1/n$. However, if a point not on $y = 0$ be chosen, and any large positive number K, it is easy to find a value of n so that $|u_n|$ exceeds K, since $\cosh(ny)$ has an exponential growth and its argument can be made large by choosing n large. Thus the solutions u_n, regarded as perturbations, are small on $y = 0$ but grow large very close to that line. Hence the solution of a boundary value problem for this equation in, say, the half-plane $y > 0$ is evidently highly unstable.

If in (1.2.5) we choose $f(z) = z^n$, $g(\bar{z}) = 0$, and separate real and imaginary parts, we have

$$z^n = (x+iy)^n = P_n(x, y)+iQ_n(x, y),$$

where P_n and Q_n are polynomials of degree n in x and y, and satisfy (1.2.4). Set $x = r\cos\theta$, $y = r\sin\theta$, so that r, θ are polar coordinates; then by de Moivre's theorem, we see that

$$P_n(x, y) = r^n\cos n\theta, \qquad Q_n(x, y) = r^n\sin n\theta.$$

Let us consider the problem of finding a harmonic function which takes given values $g(\theta)$ on the unit circle $r = 1$. Suppose $g(\theta)$ is differentiable, and has a Fourier series,

$$g(\theta) = \tfrac{1}{2}a_0+ \sum_{n=1}^{\infty} (a_n \cos n\theta + b_n \sin n\theta),$$

where $\quad a_n = \frac{1}{\pi}\int_0^{2\pi} g(\alpha)\cos n\alpha \, d\alpha, \qquad b_n = \frac{1}{\pi}\int_0^{2\pi} g(\alpha)\sin n\alpha \, d\alpha.$

Now form the series of harmonic functions

$$\tfrac{1}{2}a_0+ \sum_{n=1}^{\infty} r^n(a_n \cos n\theta + b_n \sin n\theta).$$

This series converges uniformly for $r < 1$, and its limit as $r \to 1$ is $g(\theta)$.

The sum is, therefore, a harmonic function $u(r, \theta) = u(x, y)$ with the right boundary values. If we use the formulae for a_n, b_n and change the order of summation and integration, we find

$$u(x, y) = \frac{1}{\pi} \int_0^{2\pi} g(\alpha) \left\{ \tfrac{1}{2} + \sum_{n=1}^{\infty} r^n \cos n(\theta - \alpha) \right\} d\alpha$$

$$= \frac{1}{2\pi} \int_0^{2\pi} g(\alpha) \frac{1 - r^2}{1 - 2r \cos(\theta - \alpha) + r^2} \, d\alpha. \tag{1.2.6}$$

This is the Poisson integral for the unit circle. Note that the integrand has a singularity when $r = 1$, $\theta \to \alpha$. Thus the way in which the solution depends on the boundary values is more complicated than for the wave equation. For $r < 1$, the Poisson integral is an analytic function of r and θ; hence also of x and y.

Exercise. The function

$$u(r, \theta) = \int_{-\infty}^{\infty} r^a [f(a) \cos a\theta + g(a) \sin a\theta] \, da$$

is a formal solution of

$$\frac{\partial^2 u}{\partial r^2} + \frac{1}{r} \frac{\partial u}{\partial r} + \frac{1}{r^2} \frac{\partial^2 u}{\partial \theta^2} = 0.$$

1.3. Power series solutions and existence theorems.

We shall now present two examples of the way in which solutions of analytic partial differential equations can be proved to exist, using the technique of power series expansions. The result is in each case 'local', so that the practical interest of these theorems is limited. On the other hand, the results apply to equations of very general form, and the whole method has great historical interest, since it was one of the first existence proofs for any kind of differential equation. Indeed, Cauchy, who invented the method, applied it first to ordinary differential equations. We shall need to introduce the idea of a dominant power series.

A power series $f(x) = \sum_{n=0}^{\infty} a_n x^n$ is said to be dominated by a second power series, say $F(x) = \sum_{n=0}^{\infty} A_n x^n$, provided that

$$|a_n| \leqslant A_n \quad (n = 0, 1, 2, \ldots).$$

This relation is often written

$$f(x) \ll F(x).$$

If the dominant series $F(x)$ converges for a given value of x, then certainly the dominated series also converges there. In addition, if $P(a_1,...,a_N)$ is any polynomial with positive coefficients, we see that

$$|P(a_1,...,a_N)| < P(A_1,...,A_N).$$

Thus if $f(x) \ll F(x)$, it follows that $\{f(x)\}^n \ll \{F(x)\}^n$, and so on.

It is convenient to have at hand dominant functions of the simplest possible form. Let r be any number such that the series $\sum\limits_{n=0}^{\infty} a_n r^n$ converges. Then the absolute values of the terms of this series, namely $|a_n r^n|$, have an upper bound M. It is easy to see then that the series

$$\sum_{n=0}^{\infty} M(x/r)^n = \frac{M}{1-(x/r)}$$

dominates the given series $f(x) = \sum a_n x^n$ for $|x| < r$.

Exercise. If $f(0) = 0$, then $M\{1-(x/r)\}^{-1} - M$ is a dominant.

Similar remarks hold for power series of several independent variables. If the series

$$f(x^i) = \sum a_{mn...r}(x^1)^m (x^2)^n ... (x^N)^r$$

converges for $x^i = r^i$ $(i = 1,...,N)$, then the power series

$$\varphi(x^i) = M \sum \left(\frac{x^1}{r^1}\right)^m \left(\frac{x^2}{r^2}\right)^n ... \left(\frac{x^N}{r^N}\right)^r$$

$$= M \Big/ \left\{ \left(1 - \frac{x^1}{r^1}\right)\left(1 - \frac{x^2}{r^2}\right)...\left(1 - \frac{x^N}{r^N}\right) \right\},$$

where M is chosen as in the one-variable case, provides a dominant function. Another dominant function is

$$M \Big/ \left\{ 1 - \left(\frac{x^1}{r^1} + \frac{x^2}{r^2} + ... + \frac{x^N}{r^N}\right) \right\},$$

as is easily verified by comparing coefficients with the first dominant. Other combinations of the terms in the denominator can also be made.

Exercise. Yet another dominant is

$$M \Big/ \left[\left\{ 1 - \left(\frac{x^1}{r^1} + ... + \frac{x^k}{r^k}\right) \right\} \left\{ 1 - \left(\frac{x^{k+1}}{r^{k+1}} + ... + \frac{x^N}{r^N}\right) \right\} \right].$$

Consider now a single equation of the first order which has been solved for one of the derivatives:

$$\frac{\partial z}{\partial x^1} = f\left(x^1, ..., x^N, z, \frac{\partial z}{\partial x^2}, ..., \frac{\partial z}{\partial x^N}\right),$$ (1.3.1)

and in which the function f is analytic in each of its arguments. Note that $\partial z/\partial x^1$ appears only on the left of this equation; we shall call this the normal form. Suppose that we are also given an analytic function $\varphi(x^2, ..., x^N)$ regular near some point, say the origin, on the hyperplane $x^1 = 0$. We intend to show that (1.3.1) has a unique analytic solution which reduces to $\varphi(x^2, ..., x^N)$ on the hyperplane, and which is defined in an N-dimensional neighbourhood of the origin. In short, we state the theorem: *An analytic equation of the first order in normal form has a unique analytic solution in some neighbourhood of the origin, which reduces to a given analytic function on the hyperplane $x^1 = 0$.*

Proof will be given by actual construction of a Taylor's series, centred at the origin, for the solution, and by showing that this series has a positive radius of convergence. The coefficients of this series are the values of the various derivatives of the solution at the origin, divided by certain factorials. Those derivatives which involve no differentiation with respect to x^1 may be found directly from the given values $\varphi(x^2, ..., x^N)$ on $x^1 = 0$. Those which contain one differentiation with respect to x^1 may clearly be found by differentiating (1.3.1) the necessary number of times with respect to the remaining variables, and then substituting known values in the right-hand side. Continuing in this recursive process, we may determine, as functions of $x^2, ..., x^N$ the values of all derivatives of the solution function on the hyperplane $x^1 = 0$, and in particular, the values at the origin. Now this process involves only operations of addition, multiplication, and differentiation, through all of which the relation of a dominant function is preserved. From the manner in which it was calculated, the Taylor series formally satisfies both the differential equation and the auxiliary condition on the hyperplane. It remains only to show that the series converges in some region containing the origin, and for this purpose the dominant function method will be employed.

Before we construct an actual dominant series, it is advantageous to simplify the form of the problem by taking as dependent variable a new function

$$u \equiv z - \varphi(x^2, ..., x^N),$$

which will now be required to vanish on $x^1 = 0$. The equation for u is, after substitution,

$$\frac{\partial u}{\partial x^1} = a + f_1\left(x^1, ..., x^N, u, \frac{\partial u}{\partial x^2}, ..., \frac{\partial u}{\partial x^N}\right),$$

where f_1 is a power series with zero constant term. A further reduction is possible; set $v = u - ax^1$; then

$$\frac{\partial v}{\partial x^1} = f_2\left(x^1, ..., x^N, v, \frac{\partial v}{\partial x^2}, ..., \frac{\partial v}{\partial x^N}\right),\tag{1.3.1'}$$

where f_2 is another power series with zero constant term. The auxiliary condition for v is now $v = 0$ on $x^1 = 0$.

A dominant series V for v will next be found. The function f_2 has a dominant

$$F_2 = M\Big/\left[\left\{1 - \frac{x^1 + ... + x^N + v}{r}\right\}\left\{1 - \frac{1}{\rho}\left(\frac{\partial v}{\partial x^2} + ... + \frac{\partial v}{\partial x^N}\right)\right\}\right] - M,$$

where M, r, and ρ are fixed positive constants. In order to provide for a difficulty which arises later, let us replace x^1 in this dominant by x^1/α, where $0 < \alpha < 1$. This will only increase certain coefficients in the expansion of F_2 and will therefore not destroy the domination of F_2 over f_2. Consider then the dominant differential equation

$$\frac{\partial V}{\partial x^1} = M\Big/\left[\left\{1 - \frac{1}{r}\left(\frac{x^1}{\alpha} + x^2 + ... + x^N + V\right)\right\}\left\{1 - \frac{1}{\rho}\left(\frac{\partial V}{\partial x^2} + ... + \frac{\partial V}{\partial x^N}\right)\right\}\right] - M.$$

$$\tag{1.3.2}$$

We shall try to show that (1.3.2) has a convergent series solution with only positive coefficients. Since coefficients are to be calculated by additions or multiplications of terms independent of x^1, the coefficients of any such solution are at least equal to those of any regular integral which vanishes when $x^1 = 0$. To find a solution of (1.3.2) with positive coefficients only, we may take for V a function of the single variable

$$x = \frac{x^1}{\alpha} + x^2 + ... + x^N.\tag{1.3.3}$$

After an easy calculation, we find for $V = V(x)$ the following ordinary differential equation of the second degree:

$$\left(\frac{1}{\alpha} - \frac{N-1}{\rho}M\right)\frac{dV}{dx} = \frac{N-1}{\alpha\rho}\left(\frac{dV}{dx}\right)^2 + \frac{M}{\{1 - (x + V)/r\}} - M.\tag{1.3.4}$$

Let us now choose α so small that the coefficient of dV/dx is positive. Then (1.3.4), regarded as a quadratic in dV/dx, has two distinct roots, one of which tends to zero as x and V approach zero. If we denote this root by

$\psi(x, V)$, we have reduced our problem to showing that the ordinary differential equation

$$\frac{dV}{dx} = \psi(x, V) \tag{1.3.5}$$

possesses a solution $V(x)$ with positive coefficients.

From the theory of ordinary differential equations (see Ex. 7) we may conclude that (1.3.5) has a series solution $V(x)$ which vanishes when $x = 0$. Since $\psi(0, 0) = 0$ we see that the first derivative dV/dx also vanishes for $x = 0$. Since this solution satisfies (1.3.4), and since the coefficient on the left of (1.3.4) is positive, we see that

$$\frac{dV}{dx} = a\left(\frac{dV}{dx}\right)^2 + b(x, V),$$

where a is a constant and $b(x, V)$ a series with positive terms. Thus

$$\frac{d^2V}{dx^2} = 2a\frac{dV}{dx}\frac{d^2V}{dx^2} + \frac{\partial b}{\partial x} + \frac{\partial b}{\partial V}\frac{dV}{dx};$$

since V and dV/dx vanish for $x = 0$, we see that d^2V/dx^2 is positive there. Continuing in this way we can verify that all coefficients in the expansion of $V(x)$ about $x = 0$ are positive. Thus (1.3.4), and therefore (1.3.2), has a convergent series solution with positive coefficients. Thus again, (1.3.2) has a convergent series solution which vanishes when $x^1 = 0$; and this convergent series dominates the series obtained for (1.3.1'), which therefore converges in a neighbourhood of the origin. The existence of our solution as stated in the theorem is assured. The uniqueness of this analytic solution is also established, because the construction of the Taylor series leads to a unique set of coefficients.

Let us suppose that the functions $f(x^i, z, \partial z/\partial x^i)$ and $\varphi(x^2, ..., x^N)$ are regular analytic for all values of their arguments on a region R of $x^1 = 0$ which is closed and bounded, that is, compact. Having brought the equation to the form (1.3.1) we may then find numbers M, r, and ρ such that F_2 dominates f_2 at every point of R. Supposing that the dominant series converges when each of its arguments is less than a fixed positive ϵ, the series solution centred at any point of R must converge for $|x^1| < \epsilon$. The series solutions centred at two points of R, whose neighbourhoods of convergence overlap, agree at any point of R in the overlap and also have identical values for the derivatives of all orders. Thus they represent the same real analytic function which is then a solution for $(x^2, ..., x^N)$ in R and $|x^1| < \epsilon$.

A very similar existence proof can be given for systems of partial differential equations of the first order in normal form:

$$\frac{\partial z_i}{\partial x^1} = f_i\left(x^1,...,x^N, z_1,...,z_M, \frac{\partial z_1}{\partial x^2},..., \frac{\partial z_M}{\partial x^N}\right) \qquad (i = 1,..., M).$$

In this system there are M equations for the M dependent variables $z_1,...,z_M$. Equations, or systems, of order higher than the first, are often formally equivalent to such a system.

The second of our results based on the use of dominant functions is for a rather different type of equation, or more precisely, system of equations of the first order. Let $f_i(x^1,...,x^N, z)$, $(i = 1,...,N)$, be N functions of the variables indicated. A total differential equation is a relation

$$dz = \sum_{i=1}^{N} f_i(x^k, z) \, dx^i \qquad (1.3.6)$$

among the differentials of the x^i and z. This relation is in fact equivalent to the N partial differential equations

$$\frac{\partial z}{\partial x^i} = f_i(x^k, z) \qquad (1.3.7)$$

for the dependent variable $z(x^i)$. It is not at first clear whether this system of equations is compatible, that is, whether any solutions exist at all.

Suppose, in fact, that a solution $z(x^i)$ exists. Then we may calculate the mixed second derivatives of z in two different ways: either as

$$\frac{\partial^2 z}{\partial x^i \partial x^k} = \frac{\partial}{\partial x^i} f_k(x^j, z) = \frac{\partial f_k}{\partial x^i} + \frac{\partial f_k}{\partial z} f_i$$

or with the indices i and k interchanged. Thus we must have the $\frac{1}{2}N(N+1)$ relations

$$\frac{\partial f_i}{\partial x^k} + \frac{\partial f_i}{\partial z} f_k = \frac{\partial f_k}{\partial x^i} + \frac{\partial f_k}{\partial z} f_i. \qquad (1.3.8)$$

We shall consider here only the case in which these conditions of integrability are satisfied identically in the variables x^i and z. In this case the total differential equation (1.3.6), or the system (1.3.7), is said to be completely integrable. In analytical mechanics a constraint imposed upon a mechanical system may take the form of a total differential equation. If the equation is completely integrable, the constraint is of the type known as holonomic; it can be integrated and expressed in finite terms. An example of a non-holonomic constraint is furnished by a sphere rolling on a surface—the condition of rolling without slipping is a total differential equation not completely integrable.

Relative to completely integrable equations, we have the following theorem: *A completely integrable analytic total differential equation has a unique analytic solution which takes on a given value at a given point.*

In the proof, we can surely assume that the point in question is the origin, and that the given value is zero. Now the equations (1.3.7) and those derived from (1.3.7) by differentiations enable us to calculate step by step values of the partial derivatives of all orders of z at the origin. However, the mixed partial derivatives can be computed in two or more ways, and we must show that these different ways lead to one and the same result. This is where the conditions of integrability are needed; in fact these conditions express the equality of the two values which can be found for the mixed second derivatives. Let us use induction on the order of the derivatives to show the same fact for higher orders.

In differentiating (1.3.7) with respect to the x^i, we must take account of the dependence of z on the x^i. Let us set, for any function $U(x^i, z)$,

$$\frac{dU}{dx^i} = \frac{\partial U}{\partial x^i} + \frac{\partial U}{\partial z} f_i;$$

then we have

$$\frac{\partial^2 z}{\partial x^i \partial x^k} = \frac{d}{dx^i} f_k(x^j, z), \qquad \frac{\partial^3 z}{\partial x^i \partial x^j \partial x^k} = \frac{d}{dx^i} \frac{d}{dx^j} f_k(x^j, z),$$

and it is clear that any derivative of z with respect to the x^i can be expressed as a derivative with the straight d. We must show that the order of application of these differentiations with the straight d is immaterial. Now we have

$$\frac{d^2 U}{dx^i dx^k} = \frac{d}{dx^i}\left(\frac{\partial U}{\partial x^k} + \frac{\partial U}{\partial z} f_k\right)$$

$$= \frac{\partial^2 U}{\partial x^i \partial x^k} + \frac{\partial^2 U}{\partial x^k \partial z} f_i + \frac{\partial^2 U}{\partial x^i \partial z} f_k + \frac{\partial U}{\partial z}\left(\frac{\partial f_k}{\partial x^i} + \frac{\partial f_k}{\partial z} f_i\right) + \frac{\partial^2 U}{\partial z^2} f_i f_k.$$

Of the terms in this expression, the first, the second and third together, and the last, are symmetric in i and k. In view of (1.3.8) the remaining term is also; so that the whole is symmetric and the order of differentiation does not matter. Thus the Taylor series for z centred on the origin is well-determined. Again, the coefficients of this series are obtained only by operations on the $f_i(x^k, z)$ which preserve any dominant relation.

To construct a series Z which will dominate z, let us replace the functions $f_i(x^k, z)$ in (1.3.6) by a single function $\varphi(x^j, z)$ which dominates each of the f_i and which also leads to a completely integrable total differential equation for the dominant Z.

If we take $\varphi = M\Big/\Big\{\Big(1-\dfrac{x^1+\ldots+x^N}{r}\Big)\Big(1-\dfrac{Z}{\rho}\Big)\Big\}$,

where M is sufficiently large, r and ρ sufficiently small, φ will dominate each of the $f_i(x^k, z)$, and the total differential equation for Z will be

$$dZ = \frac{M(dx^1+\ldots+dx^N)}{\{1-(x^1+\ldots+x^N)/r\}\{1-(Z/\rho)\}}. \tag{1.3.9}$$

Not only is (1.3.9) completely integrable, due to the symmetry with respect to the x^i, but solutions of it can be found by setting $x = x^1+\ldots+x^N$ whence we find for $Z = Z(x)$ the ordinary differential equation

$$\Big(1-\frac{Z}{\rho}\Big)\,dZ = \frac{M\,dx}{1-(x/r)}. \tag{1.3.10}$$

This equation, already separated, has an analytic integral $Z(x)$ which vanishes for $x = 0$ and has only positive coefficients when developed in a series of powers of x (see Ex. 6). It follows that (1.3.9) has an analytic integral vanishing when $x^1 = x^2 = \ldots = x^N = 0$, whose development in the x^i has positive coefficients only. Hence finally the total differential equation (1.3.6) has an analytic solution regular in a neighbourhood about the origin and reducing at that point to zero. This proves the theorem.

A comparison of these two results shows that where there are N equations to satisfy, it is enough to give the value at a single point, whereas, with only one equation, values can be assigned on a hyperplane of $(N-1)$ dimensions.

Exercise 1. If $f_1(x, y)$ and $f_2(x, y)$ are not both zero at a point P, and are analytic, show that the differential $f_1\,dx+f_2\,dy$ has an integrating factor μ; i.e. that $\mu(f_1\,dx+f_2\,dy) = dF$, is exact, and that an integral F exists, locally.

Exercise 2. The differential expression in three variables

$$f_1\,dx+f_2\,dy+f_3\,dz$$

has an integrating factor, if and only if

$$f_1\Big(\frac{\partial f_2}{\partial z}-\frac{\partial f_3}{\partial y}\Big)+f_2\Big(\frac{\partial f_3}{\partial x}-\frac{\partial f_1}{\partial z}\Big)+f_3\Big(\frac{\partial f_1}{\partial y}-\frac{\partial f_2}{\partial x}\Big) = 0.$$

Exercise 3. Find an integrating factor and integrate

$$(y^2+yz)\,dx+(xz+z^2)\,dy+(y^2-xy)\,dz = 0.$$

Exercise 4. Find integrals of the total differential equations

(a) $dz = \left(1 - 2\dfrac{x}{y}\right) dx + \left\{\left(\dfrac{x}{y}\right)^2 - \left(\dfrac{x}{y}\right) - \dfrac{z}{y}\right\} dy,$

(b) $dz = \dfrac{x^2 + yz}{1 - xy}\, dx + \dfrac{y^2 + xz}{1 - xy}\, dy.$

Exercise 5. Find $f(y)$, on the assumption that

$$dz = f(y)[dx - z\, dy]$$

is completely integrable; and integrate the resulting equation.

Exercise 6. Show explicitly that (1.3.10) has an analytic solution whose series expansion contains only positive coefficients, by integrating the equation.

Exercise 7. Show that any analytic ordinary differential equation of the form (1.3.5) has analytic solutions vanishing at $x = 0$ by showing that a dominant can always be found which satisfies an equation of the form (1.3.10).

1.4. Transformations of variables ; tensors. Many of the partial differential equations which have been studied by pure or applied mathematicians have arisen directly from geometrical or physical problems. Indeed, such an equation is usually the statement of some geometric fact or physical law. However, in order to express this fact or law in terms of a differential equation, it is necessary to adopt a coordinate system. The coordinates so chosen will usually make their appearance in the partial differential equation of the problem as independent variables. Now we might choose two systems of coordinates and write down the differential equation in each of these two systems. The two equations, although formally rather different, are in reality one and the same, since to each solution of one corresponds a solution of the other, and under the appropriate change of variables one equation transforms into the other. It is therefore desirable to regard the two forms of the equation as the same, and to employ a notation such that the equation would have the same *form* in all coordinate systems, that is, would be covariant.

Such a formalism is provided by the tensor calculus, and we summarize some basic definitions necessary in this calculus. We have already mentioned the concept of the space of the independent variables x^i, and of the points P of that space. In a geometrical problem this space will frequently coincide with the 'geometrical space'; in a physical problem the space of the independent variables may be the 'physical space' itself, or it may be a 'configuration space', of which each point corresponds to a configuration of the physical system under study. The independent variables x^i may be called coordinates in this space. A curve is defined by N functions $x^i(t)$ of a single parameter; a hypersurface by a single equation $F(x^1,..., x^N) = 0$.

For economy of writing we shall use the two following conventions: (a) the range convention—a small latin suffix unrepeated in a term is understood to take in succession all values from 1 to N inclusive, (b) the summation convention—a small latin suffix repeated in a term is understood to take all values $1,..., N$, and the N terms so formed are *added together*.

A transformation of coordinates (a change of independent variables) is given by a set of N equations

$$x'^r = f^r(x^s),\qquad (1.4.1)$$

where the f^r are suitably defined real-valued functions on the space. To each point P, these equations assign a set of numerical values of the new coordinates x'^r. We shall always assume that the functional determinant of the transformation does not vanish:

$$J \equiv \frac{\partial(x'^r)}{\partial(x^s)} \equiv \frac{\partial(f^1,...,f^N)}{\partial(x^1,..., x^N)} \neq 0,\qquad (1.4.2)$$

so that (1.4.1) is a non-degenerate transformation. Thus in particular we can solve (1.4.1) for the x^s:

$$x^s = g^s(x'^r);\qquad (1.4.3)$$

these are the equations of the inverse transformation.

If now we differentiate (1.4.1), we have

$$dx'^r = \frac{\partial x'^r}{\partial x^s}\, dx^s.\qquad (1.4.4)$$

Here we have made use of the range and summation conventions. The components dx^r represent an *infinitesimal displacement*; we see from (1.4.4) the law of transformation of these components under a transformation of coordinates. An infinitesimal displacement is the prototype and most important example of a *contravariant vector*. A set of N quantities T^r, associated with a point P and each coordinate system at P, and which transform with the coordinate system according to the law

$$T'^r = \frac{\partial x'^r}{\partial x^s}\, T^s,\qquad (1.4.5)$$

are said to be components of a contravariant vector at P. When no confusion is likely, we can drop any mention of the point P. Note that (1.4.5) is a homogeneous linear transformation, so that if $T^r = 0$ in one coordinate system then T^r vanishes in all coordinate systems.

A contravariant vector is an example of a *tensor*. The 'simplest possible' tensor is an *invariant*, that is, a point function which does not change in value under coordinate transformations. If $\varphi = \varphi(P) = \varphi(x^1,...,x^N)$ is such an invariant, then the partial derivatives of φ transform under coordinate transformations according to the chain rule formula

$$\frac{\partial \varphi}{\partial x'^r} = \frac{\partial x^s}{\partial x'^r} \frac{\partial \varphi}{\partial x^s}. \tag{1.4.6}$$

The *gradient vector* $\partial \varphi / \partial x^r$ is the prototype of a class of tensors known as *covariant vectors*. The general definition of a covariant vector is the same as that of a contravariant vector except that the suffix is written as a subscript and the transformation law is

$$T'_r = \frac{\partial x^s}{\partial x'^r} T_s. \tag{1.4.7}$$

Geometrically, a covariant vector represents a surface element, though a covariant vector field is not necessarily the gradient vector of an invariant scalar.

Thus a contravariant vector T^r at P determines the 'direction' of an infinitesimal displacement $T^r dt$; while a covariant vector T_r gives the orientation at P of a surface element dS, where, if S has the equation $F(x^r) = 0$, the components of the gradient vector $\partial F / \partial x^r$ are proportional at P to T_r. Contravariant and covariant vectors are the two kinds of tensors of the first order.

Tensors of higher order are defined by more elaborate laws of transformation analogous to (1.4.5) and (1.4.7). For instance, a set of N^2 quantities T^{rs} which transform according to the rule

$$T'^{rs} = \frac{\partial x'^r}{\partial x^m} \frac{\partial x'^s}{\partial x^n} T^{mn}, \tag{1.4.8}$$

are components of a contravariant tensor of the second order. Quantities $T^r{}_s$ such that

$$T'^r{}_s = \frac{\partial x'^r}{\partial x^m} \frac{\partial x^n}{\partial x'^s} T^m{}_n, \tag{1.4.9}$$

are the components of a mixed tensor of the second order. All such transformation rules are linear and homogeneous, so that the sum of two tensors of a certain kind is again a tensor of the same kind. Also, if the components of any tensor vanish in any one coordinate system, they vanish in all; the tensor is then the zero tensor of that type. An 'outer product' $A^r B^s$ of two contravariant vectors is a contravariant tensor of order two, while, for example, $A^r C_s$ is a mixed tensor. On account of the chain rule for partial derivatives, the contraction (in this case a scalar product) $A^r C_r$ of a contravariant and a covariant vector yields an invariant, for

$$A'^r C'_r = A^s \frac{\partial x'^r}{\partial x^s} \frac{\partial x^m}{\partial x'^r} C_m = A^s C_s,$$

since

$$\frac{\partial x'^r}{\partial x^s} \frac{\partial x^m}{\partial x'^r} = \delta^m_s,$$

where δ^m_s, the Kronecker delta, is zero if $m \neq s$, unity if $m = s$. We leave the reader to prove that δ^m_s is actually a mixed tensor; that is, it satisfies (1.4.9). The Kronecker delta δ^m_s can be regarded as a kind of 'unit matrix', or as a substitution operator which substitutes one of its indices for the other; it is one of the few 'numerical' tensors.

This brief sketch of coordinate transformations and tensors is sufficient for our immediate needs in treating partial differential equations of the first order. Some exercises on this topic are included below; for a more thorough treatment of these the reader may consult (33). Later in this book we shall introduce the ideas of Riemannian geometry—such as length, angle, volume—which play an essential role in connexion with linear partial differential equations of the second order.

Exercise 1. Write down transformation laws for

(a) a second-order covariant tensor T_{rs};

(b) a mixed tensor T^{rs}_m.

Exercise 2. If T^{rs} is symmetric ($T^{rs} = T^{sr}$) in one coordinate system, it is symmetric in all coordinate systems. Similarly for skew-symmetry of T^{rs}: $T^{rs} = -T^{sr}$.

Exercise 3. If $A^r X_r$ is an invariant, for every covariant X_r, then A^r must be a contravariant vector.

Exercise 4. If $a_{rs} X^r Y^s$ is an invariant for every pair of contravariant vectors X^s, Y^s, then a_{rs} is a covariant tensor.

II

LINEAR EQUATIONS OF THE FIRST ORDER

PARTIAL differential equations of the first order are of great importance in diverse applications: for instance, in differential geometry and in analytical mechanics, as well as in the further development of integration theories for equations of higher order. Here we shall study linear equations, postponing to the next chapter the more elaborate treatment necessary for equations of general non-linear form. First we consider a single linear homogeneous equation, interpret its geometric meaning, and show that the integration can be reduced, in theory, to the integration of ordinary differential equations. Equations of a type known as quasi-linear can be transformed to linear equations; this is shown in the second section. The solution of initial value problems for these equations may be achieved by a suitable choice of certain arbitrary functions which appear in the general solution. In simple cases, this choice can be made with surprising ease.

The remainder of the chapter is centred on the study of systems of linear homogeneous equations, the methods used being suitable modifications of those already developed. The geometrical meaning of such a system leads naturally to the concept of an 'adjoint system' of total differential equations, which, for a single equation, reduces to the ordinary differential equations previously studied.

2.1. Homogeneous linear equations. In studying linear equations of the first order it is simplest to begin with the homogeneous linear equation, which can be given a most direct geometrical interpretation. The tensor notation introduced in the preceding chapter makes it just as easy to work with N independent variables as with two, so we shall consider the general case of N variables forthwith. The homogeneous linear equation is

$$b^i \frac{\partial u}{\partial x^i} = 0, \qquad (2.1.1)$$

where the b^i are N given functions of the independent variables x^j. We shall suppose that these functions are differentiable and that not all of them vanish simultaneously. The nature of the condition imposed on the dependent variable $u(x^j)$ is clearly unaltered if the b^i are multiplied by a non-vanishing common multiplier.

Under a transformation of variables,

$$x'^i = f^i(x^j), \tag{2.1.2}$$

we have

$$\frac{\partial u}{\partial x^i} = \frac{\partial x'^j}{\partial x^i}\frac{\partial u}{\partial x'^j}$$

so that (2.1.1) becomes

$$b'^j\frac{\partial u}{\partial x'^j} = 0,$$

where

$$b'^j = \frac{\partial x'^j}{\partial x^i}b^i.$$

We may therefore assume that the N functions b^i transform like the components of a contravariant vector; the notation has anticipated this result.

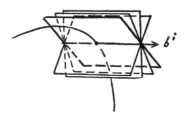

FIG. 1. Integral surface elements of the linear
equation. One of them touches the initial
manifold (shown as a curve).

The equation itself is always given in a particular coordinate system, so that the components b^i in that system are given functions of those coordinates. We shall call b^i the *characteristic* vector. It is seen that the 'length' or magnitude of this vector may be regarded as undetermined but that its direction, which we may call the *characteristic direction*, is uniquely determined by the ratios of the coefficients in the differential equation.

Since b^i is a contravariant vector, the characteristic direction is specified independently of coordinate transformations, and the infinitesimal displacement

$$dx^i = b^i\, dt \tag{2.1.3}$$

is defined at each point. Now suppose that $u(x^i)$ is a solution function of the differential equation. Under the displacement (2.1.3) we have

$$du = \frac{\partial u}{\partial x^i}\, dx^i = b^i\frac{\partial u}{\partial x^i}\, dt = 0. \tag{2.1.4}$$

Thus the numerical value of u is unchanged. Geometrically, this means that the displacement (2.1.3) is tangent to the level surface $u(x^i) = $ constant (Fig. 1).

An integral surface S of (2.1.1) we shall define as a level surface $u(x^i) = $ constant, where $u(x^i)$ is any solution function. Therefore the differential equation has the geometrical interpretation that the characteristic direction is tangent to every integral surface. That is, every integral surface S contains as a tangential direction the fixed characteristic direction. This suggests that we should construct the family of curves whose tangent at each point is the characteristic direction. By definition, then, these curves, known as characteristic curves, satisfy the system of ordinary differential equations

$$\frac{dx^i}{dt} = b^i. \tag{2.1.3'}$$

The auxiliary parameter t, which now appears as independent variable, is determined up to an additive constant once the vector b^i is given. From the theory of ordinary differential equations we assume that (2.1.3) has a family of solutions, of the form $x^i = X^i(t)$, which represent integral curves of the characteristic vector field b^i (18). Through each point of space passes one and only one of these integral or 'characteristic' curves. From (2.1.4) we see that, along a characteristic curve, any solution function u is a constant. Therefore, an integral surface $S:u = $ constant, on which lies one point of a characteristic curve, must contain the entire curve. For example, if two integral surfaces meet in a point, they must meet in the whole length of the characteristic curve through that point. Indeed, if two integral surfaces are in addition tangent at a point, they must also be tangent along the entire characteristic curve through the point.

Conversely, we may observe that every integral surface is constructed, so to speak, of characteristic curves. For through each point of the integral surface there passes an integral curve of (2.1.3), that is to say, a characteristic curve, and this curve must lie entirely within the integral surface. Since the integral surface is of $(N-1)$ dimensions, there is an $(N-2)$-parameter family of characteristic curves lying on the surface, one passing through each point of the integral surface.

If the parameter t is omitted, the equations (2.1.3) may be written

$$\frac{dx^1}{b^1} = \frac{dx^2}{b^2} = \dots = \frac{dx^N}{b^N},$$

a set of $(N-1)$ equations among the variables x^i. The integral curves of this system will appear as $(N-1)$ functionally independent finite equations

$$f_{(\alpha)}(x^j) = C_\alpha \ (\alpha = 1,\dots, N-1). \tag{2.1.5}$$

Any one equation of the form (2.1.5) which holds as a consequence of

(2.1.3) is known as a first integral of the system of ordinary differential equations (2.1.3). It is clear that the form of the equations (2.1.5) is by no means uniquely determined, since we can replace the $(N-1)$ functions $f_{(\omega)}(x^j)$ by any $(N-1)$ other functions $F_{(\beta)}(f_{(\omega)})$ which are independent functions of the $f_{(\omega)}(x^j)$. When numerical values of the constants C_α are assigned, (2.1.5) represents a particular characteristic curve which might, for example, be so chosen as to pass through a given point. When the values of the C_α are left undetermined, we speak of (2.1.5) as the general integral of (2.1.3).

No matter how this general integral is expressed, the functions $f_{(\omega)}(x^j)$ all satisfy our original partial differential equation (2.1.1). For, upon differentiating (2.1.5) along a characteristic curve and using (2.1.3), we obtain

$$0 = df_{(\omega)} = \frac{\partial f_{(\omega)}}{\partial x^j}\, dx^j = \frac{\partial f_{(\omega)}}{\partial x^j} b^j\, dt,$$

and it follows that $b^i \dfrac{\partial f_{(\omega)}}{\partial x^i} = 0$, as stated. The $f_{(\omega)}(x^j)$ therefore constitute $(N-1)$ functionally independent integrals of the partial differential equation. Moreover, any function $\pi(f_{(\omega)})$ of these integrals is also an integral. To prove this, we need only observe that

$$b^i \frac{\partial \pi}{\partial x^i} = \sum_\alpha b^i \frac{\partial \pi}{\partial f_{(\omega)}} \frac{\partial f_{(\omega)}}{\partial x^i} = \sum_\alpha \frac{\partial \pi}{\partial f_{(\omega)}} b^i \frac{\partial f_{(\omega)}}{\partial x^i} = 0. \qquad (2.1.6)$$

(Since α is not a tensor index, but an index of enumeration, the summation over α is indicated explicitly.)

Now we can show, conversely, that all integrals of the partial differential equation can be obtained this way. Suppose, in fact, that $f(x^j)$ is any integral whatever of the differential equation (2.1.1). Then the N homogeneous linear equations

$$b^i \frac{\partial f}{\partial x^i} = 0,$$

$$b^i \frac{\partial f_{(\omega)}}{\partial x^i} = 0 \qquad (2.1.7)$$

have a solution b^i not identically zero. It follows that the functional determinant

$$\frac{\partial(f, f_{(1)}, \ldots, f_{(N-1)})}{\partial(x^1, x^2, \ldots, x^N)}$$

vanishes identically. Therefore the integral f and the $(N-1)$ integrals $f_{(\omega)}$ are functionally dependent, as was to be shown.

Let us state the results just demonstrated in the form of a theorem: If $f_{(\omega)}(x^j) = C_\alpha$, $\alpha = 1,..., (N-1)$, are $N-1$ functionally independent first integrals of the characteristic equations

$$\frac{dx^1}{b^1} = \frac{dx^2}{b^2} = ... = \frac{dx^N}{b^N},$$

then every function $\pi(f_{(\omega)})$ of the $f_{(\omega)}$ is an integral of the linear partial differential equation

$$b^i \frac{\partial U}{\partial x^i} = 0.$$

Conversely, every integral of the latter equation is of the form $\pi(f_{(\omega)})$.

Example. Let the equation be

$$x\frac{\partial u}{\partial x} + y\frac{\partial u}{\partial y} + z\frac{\partial u}{\partial z} = 0.$$

The characteristic equations are

$$\frac{dx}{x} = \frac{dy}{y} = \frac{dz}{z};$$

and have the first integrals $y = ax$, $z = bx$. Therefore the generál integral is

$$u = \pi\left(\frac{y}{x}, \frac{z}{x}\right),$$

where π is an arbitrary function of its two arguments. Suppose that it is required to find the equation of that integral surface which passes through the parabola $z^2 = 4py$ lying in the plane $x = 1$. If $x = 1$, the equation $z^2 = 4py$ implies that $b^2 = 4pa$, so that the integral surface is

$$\left(\frac{z}{x}\right)^2 = 4p\left(\frac{y}{x}\right), \quad \text{or} \quad z^2 = 4pxy.$$

On this surface, the dependent variable has a constant value. Thus, any solution u of the equation which vanishes on the given parabola has the form

$$u = f\left\{\left(\frac{z}{x}\right)^2 - 4p\left(\frac{y}{x}\right)\right\},$$

$f(t)$ being any function such that $f(0) = 0$.

The complete manifold of solutions of the given partial differential equation may therefore be characterized if we can find $N-1$ independent first integrals of the auxiliary ordinary differential equations (2.1.3). However, even if we can find only p independent first integrals, where $p < N-1$, we can at least reduce the given partial differential equation to an equation in $N-p$ variables. To verify this, let us suppose that $f_{(\alpha)}$ ($\alpha = 1,...,p$) are the first integrals in question. Since they are independent, we may choose a new coordinate system x'^j in which the p variables $x'^1,..., x'^p$ are identical with the known first integrals $f_{(1)},..., f_{(p)}$. The partial differential equation goes over into a new equation

$$b'^i \frac{\partial u}{\partial x'^i} = 0, \tag{2.1.8}$$

which has as these first integrals $u = x'^\alpha$ ($\alpha = 1,...,p$). It follows that $b'^\alpha = 0$ ($\alpha = 1,...,p$), so that (2.1.8) is in reality an equation in the remaining $N-p$ variables only. The dependent variable u may still involve, as parameters, the p variables which have been eliminated from the differential equation. Thus the discovery of each new first integral enables us to reduce the number of independent variables, and so constitutes a step toward the complete solution.

Exercise 1. Find the general solutions of

(a) $(mz-ny)\dfrac{\partial u}{\partial x} + (nx-lz)\dfrac{\partial u}{\partial y} + (ly-mx)\dfrac{\partial u}{\partial z} = 0,$

(b) $\cos x \cos y \dfrac{\partial u}{\partial x} - \sin x \sin y \dfrac{\partial u}{\partial y} + \sin x \cos y \dfrac{\partial u}{\partial z} = 0,$

(c) $yz\dfrac{\partial u}{\partial x} + zx\dfrac{\partial u}{\partial y} + xy\dfrac{\partial u}{\partial z} + xyz = 0.$

Exercise 2. Show that a first integral can be found if a covariant vector μ_i is known, such that $\mu_i b^i \equiv 0$ and $\mu_i dx^i$ is an exact differential.

Exercise 3. The most general linear equation having the $N-1$ first integrals $f_{(\alpha)}(x^j)$ is of the form

$$A(x^j) \frac{\partial(u, f_{(1)},..., f_{(N-1)})}{\partial(x^1,..., x^N)} = 0,$$

where $A(x^j)$ is an arbitrary function.

2.2. The quasi-linear equation of the first order. The equation

$$b^i \frac{\partial z}{\partial x^i} = b, \tag{2.2.1}$$

where b^i and b are functions of x^i and of the dependent variable z, is said to be quasi-linear, since the first derivatives appear linearly in it. By means of a device often used in the theory of partial differential equations of the first order, the quasi-linear equation (2.2.1) can in principle be reduced to the homogeneous linear equation of the preceding section. Let us try to define z as a function of the x^i by means of an implicit equation of the form

$$u(x^i, z) = 0. \tag{2.2.2}$$

From (2.2.2) we find, on differentiating,

$$\frac{\partial u}{\partial x^i} + \frac{\partial u}{\partial z} \frac{\partial z}{\partial x^i} = 0, \tag{2.2.3}$$

and, if we solve for $\partial z / \partial x^i$ and substitute the values so obtained in (2.2.1) we see that u satisfies the strictly linear homogeneous equation

$$b^i \frac{\partial u}{\partial x^i} + b \frac{\partial u}{\partial z} = 0. \tag{2.2.4}$$

The number of independent variables has increased by one, since we now consider z as an independent variable $z = x^{N+1}$. In the $(N+1)$-dimensional space V_{N+1} of the x^i $\{i = 1,..., (N+1)\}$, the N quantities b^i together with the function b now transform as components of a contravariant vector. For the purpose of numerical computation, the 'reduction' just given is not likely to be useful, since the amount of calculation increases rapidly with the number of variables of the problem. Indeed, a computer would rather invert this process, to transform a linear equation in three variables, say, into a quasi-linear equation in two variables, if the solution of his problem could be found in this way.

The theorem of the last section, as applied to the quasi-linear equation, therefore reads as follows: *If $f_{(\alpha)}(x^i) = $ constant $(\alpha = 1,..., N)$ are N independent first integrals of the auxiliary equations*

$$\frac{dx^1}{b^1} = ... = \frac{dx^N}{b^N} = \frac{dz}{b}, \tag{2.2.5}$$

then every function z implicitly defined by an equation

$$\pi\{f_{(\alpha)}(x^i, z)\} = 0, \tag{2.2.6}$$

where π is an arbitrary differentiable function, is a solution of the equation

$$b^i \frac{\partial z}{\partial x^i} = b. \tag{2.2.1}$$

This relation (2.2.6) is again known as the general integral. In certain exceptional cases there are integrals of the differential equation which cannot be obtained in this way, but we shall not investigate these.

Exercise. Find the general integral of

$$\frac{\partial z}{\partial x}+x\frac{\partial z}{\partial y}+xy\frac{\partial z}{\partial v}=xyvz^{\sharp}.$$

In the space V_{N+1} an integral surface S, represented by the implicit equation $u(x^i,z)=0$ is in general an N-dimensional hypersurface. Through each point of S passes a characteristic curve which is an integral curve of the differential equations (2.2.5); indeed, the characteristic curve lies entirely in S, by the reasoning of the previous section. In general, therefore, S is composed of an $(N-1)$-parameter family of characteristic curves. It should thus be possible to find an $(N-1)$-dimensional hypersurface of V_{N+1} such that each of these characteristic curves meets the hypersurface in precisely one point (Fig. 2).

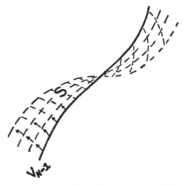

FIG. 2. Integral surface generated by characteristic curves.

Conversely, therefore, we may propose the following initial value problem: given a hypersurface V_{N-1}, let us say in the form of parametric equations

$$x^i=f^i(t^1,...,t^{N-1}),\qquad z=f^{N+1}(t^1,...,t^{N-1}),\qquad(2.2.7)$$

with $N-1$ parameters $t^{(\alpha)}$, construct an integral surface $S=V_N$, of the equation (2.2.1), which contains the given V_{N-1}. To solve this problem we first determine the characteristic curves through the points of V_{N-1} by integrating the equations (2.2.5). In this way we obtain a set of $(N+1)$ functions

$$x^i=f^i(t,t^1,...,t^{N-1}),\qquad z=f^{N+1}(t,t^1,...,t^{N-1}),\qquad(2.2.8)$$

which are continuously differentiable in each argument and which reduce to (2.2.7) when $t=0$.

To construct the actual solution z as a function of the independent variables x^i, we solve the first N equations of (2.2.8) for $t, t^1,..., t^{N-1}$ in terms of the x^i. Substituting the functions of x^i so determined in the last equation of (2.2.8), we obtain z as a function of the independent variables. The characteristic curves (2.2.8) generate a surface S in V_{N+1} which is, in general, N-dimensional (Fig. 2). The relation $z=f(x^i)$ obtained by the elimination of the N parameters $t, t^1,..., t^{N-1}$ is the equation of this surface S. We will show that $z=f(x^i)$ so determined does satisfy the differential equation, but first let us observe that we can solve for the N quantities if and only if the functional determinant

$$\Delta(t) = \frac{\partial(x^1,\ldots\ldots,x^N)}{\partial(t,t^1,\ldots,t^{N-1})} = \left| \frac{\partial x^i}{\partial t}, \frac{\partial x^i}{\partial t^\alpha} \right| \qquad (2.2.9)$$

does not vanish. Now in view of the characteristic equations (2.2.5), we have, differentiating (2.2.8) with respect to t,

$$\frac{\partial x^i}{\partial t} = b^i, \qquad (2.2.10)$$

and therefore $\qquad \Delta(0) = \left| b^i, \frac{\partial x^i}{\partial t^\alpha} \right|. \qquad (2.2.11)$

On the given surface V_{N-1}, the partial derivatives $\partial x^i/\partial t^\alpha$ and b^i are known. We shall assume for the moment that $\Delta(0) \neq 0$ on V_{N-1}. Since $\Delta(t)$ is a continuous function of t, $\Delta(t)$ will be different from zero in some neighbourhood of the initial surface, and the elimination is valid there. We may remark that 'theoretical' elimination processes of this sort are, by their very nature, applicable only locally, so that the existence of solutions can be guaranteed only in some small neighbourhood of the initial surface. This is one reason for the difficulty of proving global results. In particular cases, it is sometimes possible to carry out such eliminations explicitly, and thus show that solutions exist in the large.

Under the assumption, then, that $\Delta \neq 0$ on V_{N-1}, we can solve for the t^α and express z as a function of the x^i from (2.2.8). We must now verify that z is a solution of the partial differential equation, and for this purpose we note the equations (2.2.5). We have, in fact,

$$b^i \frac{\partial z}{\partial x^i} = \frac{\partial z}{\partial x^i} \frac{dx^i}{dt} = \frac{dz}{dt} = b; \qquad (2.2.12)$$

and this is just the statement required. Looking back over the process of solution, and recalling the standard uniqueness theorems for the solutions of ordinary differential equations (18, ch. ii) and implicit functions (12a, ch. ii), we see that the solution function z is uniquely determined in the neighbourhood where it is defined, at any rate if the coefficients b^i are differentiable. To sum up, then, *if* $\Delta \neq 0$ *on* V_{N-1}, *there exists a unique local solution of the initial value problem.*

Geometrically, the condition $\Delta \neq 0$ means that the vector b^i ($i = 1,\ldots,N$) is not tangent to the projection of V_{N-1} on the hyperplane $z = 0$, which is the space of the variables $x^1,\ldots, x^N$. The exceptional case $\Delta = 0$ must be treated differently. We shall for simplicity now suppose that $\Delta \equiv 0$ on

V_{N-1}. That is, the vector b^i shall be everywhere tangent to the projection of V_{N-1} on the hyperplane $z = 0$. Analytically, this may be represented by the fact that there exist $N-1$ well-determined functions $\lambda_{(\alpha)}(t^1, ..., t^{N-1})$ such that on V_{N-1},

$$b^i = \sum_\alpha \lambda_{(\alpha)} \frac{\partial x^i}{\partial t^\alpha}. \qquad (2.2.13)$$

Exercise. Use the fact that $\Delta = 0$, and the fact that the t^α are independent parameters on V_{N-1}, to show (2.2.13).

We shall now make the definition: V_{N-1} is *characteristic* if the characteristic direction (b^i, b) in V_{N+1} lies in V_{N-1}. That is, the characteristic vector (b^i, b) is to be a linear sum of the $(N-1)$ tangential vectors

$$\left(\frac{\partial x^i}{\partial t^\alpha}, \frac{\partial z}{\partial t^\alpha} \right) \qquad (\alpha = 1, ..., N-1).$$

If this is true, we have N equations of the form (2.2.13), together with the additional condition

$$b = \sum_\alpha \lambda_{(\alpha)} \frac{\partial z}{\partial t^\alpha}. \qquad (2.2.14)$$

It is geometrically evident that every $(N-2)$-parameter family of characteristic curves generates a characteristic manifold, and, conversely, that every characteristic manifold is generated by such an $(N-2)$-parameter family of characteristic curves.

Exercise. Give analytical proofs of these two statements. To prove the second part, consider the curves on V_{N-1}, defined by the ordinary differential equations

$$\frac{dt^\alpha}{ds} = \lambda_{(\alpha)}(t^1, ..., t^{N-1}).$$

We can now show that the initial value problem when $\Delta \equiv 0$ on V_{N-1} is solvable (and then in infinitely many ways), if and only if V_{N-1} is a characteristic manifold. From the hypothesis $\Delta \equiv 0$ it follows that $N-1$ quantities $\lambda_{(\alpha)}$ exist, such that (2.2.13) hold. Suppose now that a solution z of the equation (2.2.1) exists, then

$$b = b^i \frac{\partial z}{\partial x^i} = \sum_\alpha \lambda_\alpha \frac{\partial x^i}{\partial t^\alpha} \frac{\partial z}{\partial x^i} = \sum_\alpha \lambda_\alpha \frac{\partial z}{\partial t^\alpha},$$

the first equality being the statement that z is a solution. But this yields the further condition (2.2.14) which shows that V_{N-1} must be characteristic. Conversely, if V_{N-1} is characteristic, we can construct an integral surface from any $(N-1)$-parameter family of characteristic curves containing the $(N-2)$-parameter family which constitute V_{N-1}. Clearly this can be achieved in infinitely many ways.

It may appear that the case when $\Delta \equiv 0$ and V_{N-1} is not a characteristic manifold is left in doubt. Geometrically, this occurs when the vector (b^i, b) in V_{N+1} is not tangential to V_{N-1}, but its projection b^i on the hyperplane $z = 0$ is tangential to the projection of V_{N-1} on $z = 0$. In order to examine this case, let us reduce (2.2.1) to the linear equation (2.2.4) by an implicit change of variables as in (2.2.3). In (2.2.4), the x^i and z appear on an equal footing, so that we may choose any one of these $N+1$ variables as a new dependent variable and obtain for it a quasi-linear equation similar to (2.2.1), but with certain coefficients interchanged. It is then evident that if projections of the vector (b^i, b) are tangential to projections of V_{N-1} on all coordinate planes, then (b^i, b) must be tangent to V_{N-1} in the large space V_{N+1}, so that V_{N-1} is in fact a characteristic manifold.

Exercise. How does the preceding work simplify for the homogeneous linear equation, when b^i depends only on the x^i, and $b = 0$?

To illustrate the theory of the linear and quasi-linear equations of the first order, let us consider the quasi-linear equation in two independent variables x and y,

$$Pp + Qq = R,$$

where

$$p = \frac{\partial z}{\partial x}, \qquad q = \frac{\partial z}{\partial y},$$

and P, Q, R are functions of x, y, and z. If we attempt to define z by an implicit equation $u(x, y, z) = 0$, we obtain

$$u_x + pu_z = 0, \qquad u_y + qu_z = 0,$$

whence

$$Pu_x + Qu_y + Ru_z = 0$$

is the transformed linear differential equation. The characteristic curves have the equations

$$\frac{dx}{P} = \frac{dy}{Q} = \frac{dz}{R}.$$

If we can discover two independent first integrals $f(x, y, z) = a$, $g(x, y, z) = b$, then any relation

$$\pi(a, b) = 0$$

leads to an integral surface whose equation is $\pi(f, g) = 0$. Consider for example the equation

$$\frac{\partial z}{\partial x} + z^2 \frac{\partial z}{\partial y} = 1,$$

for which the characteristic equations are

$$\frac{dx}{1} = \frac{dy}{z^2} = \frac{dz}{1} = ds,$$

so that

$$x = x_0 + s,$$

$$z = z_0 + s,$$

and

$$y = y_0 + \frac{(z_0 + s)^3 - z_0^3}{3}.$$

These equations represent the characteristic curve through the point (x_0, y_0, z_0) in the three-dimensional space. Thus if an initial curve

$$C : x = f(t), \ y = g(t), \ z = h(t)$$

with parameter t be chosen, we have as surface generated by the characteristic curves passing through C,

$$x = f(t) + s,$$

$$y = g(t) + \tfrac{1}{3}s\{s^2 + 3sh(t) + 3h^2(t)\},$$

$$z = h(t) + s.$$

If the initial curve is nowhere characteristic, the parameters s, t can be eliminated from these three equations, resulting in an equation of the integral surface in a neighbourhood of C. If C is characteristic, as is the curve $x = t$, $y = \tfrac{1}{3}t^3$, $z = t$, we find $x = (t+s)$, $y = \tfrac{1}{3}(t+s)^3$, $z = (t+s)$, that is, the curve C over again.

Exercise. Show that the above equation has the general integral

$$y = \tfrac{1}{3}z^3 + f(z-x),$$

where f is an arbitrary function.

If we introduce Euclidean geometry in the space of the variables x, y, and z, the normal to a surface can be defined. Suppose the surface has the equation $z = f(x, y)$; the direction cosines of the normal are then proportional to p, q, and -1. The equation $Pp + Qq = R$ states, therefore, that the normal is perpendicular to the direction with direction numbers (P, Q, R); that is, the latter direction is tangential to the surface. As we have seen, however, it is not necessary to use the idea of a normal to a surface to give geometric significance to the partial differential equation; in fact there is no mention, in the equation, of any 'geometrical object' except the direction of the vector (P, Q, R), or in the general case (b^i, b).

Exercise 1. Find the differential equation of the surfaces which cut orthogonally at each point the surfaces of a given family $F(x, y, z) = C$. Integrate the equation for the case $F = ax^2 + by^2 + cz^2$.

Exercise 2. Integrate $f(y)p + g(x)q = 1$, defining such functions related to f and g as are necessary.

Exercise 3. Find the solution z of

$$axp + byq = cz,$$

which has the value unity on the circle $x^2 + y^2 = r^2$.

Exercise 4. Let $f_{(\alpha)}(x^1, ..., x^N, z)$ be $N-1$ functions ($\alpha = 1, ..., N-1$) of (x^i, z). The equation

$$\frac{\partial(z, f_{(1)}, ..., f_{(N-1)})}{\partial(x^1, ..., x^N)} = 0,$$

may be interpreted as a quasi-linear partial differential equation of the first order with z as dependent variable. Show that the general integral of this equation is to be found by setting $F(z, f_{(1)}, ..., f_{(N-1)}) = 0$, where F is an arbitrary function.

2.3. Systems of linear homogeneous equations.

Consider a system of p linear homogeneous partial differential equations in one dependent variable and N independent variables. Such a system is determined if we are given p contravariant vector fields b_ρ^i ($\rho = 1, ..., p$), for we may write the equations as follows:

$$b_\rho^i \frac{\partial u}{\partial x^i} \equiv X_\rho(u) = 0. \qquad (2.3.1)$$

We shall suppose that the p contravariant vectors b_ρ^i define independent displacements (are linearly independent at each point), so that the p vectors span a subspace of p dimensions. Then the equations (2.3.1) are said to be independent. The summation convention for Greek indices shall apply.

Exercise. The only integral of N independent equations (2.3.1) is $u = \text{constant}$.

The problem before us is to find all the common solutions $u(x^i)$. Now each of the equations (2.3.1) has the geometrical meaning described in § 2.1, namely, that the integral surface $S : u = \text{constant}$ at each point contains the direction given by the vector b_ρ^i. Therefore the integral surface contains the p independent directions b_ρ^i ($\rho = 1, ..., p$), so a displacement along any one of these vectors lies in the surface. Consider a displacement along b_ρ^i followed by displacements along b_σ^i, and then in the reverse direction along b_ρ^i and b_σ^i again. Now the displacement resulting from this path (which is not necessarily a closed circuit) lies in S. Analytically, this means that as a consequence of (2.3.1) we have the further equations

$$[X_\sigma X_\rho - X_\rho X_\sigma](u) = b_\sigma^i \frac{\partial}{\partial x^i}\left(b_\rho^j \frac{\partial u}{\partial x^j}\right) - b_\rho^i \frac{\partial}{\partial x^i}\left(b_\sigma^j \frac{\partial u}{\partial x^j}\right)$$

$$= \left(b_\sigma^i \frac{\partial b_\rho^j}{\partial x^i} - b_\rho^i \frac{\partial b_\sigma^j}{\partial x^i}\right)\frac{\partial u}{\partial x^j} = 0. \tag{2.3.2}$$

These are linear, homogeneous equations of the first order. Now the set of equations (2.3.2) may include one or more equations which are independent of (2.3.1). Let us adjoin all such new equations to (2.3.1), obtaining a new system

$$X_\rho(U) = 0 \quad (\rho = 1,...,p+p_1), \tag{2.3.3}$$

which includes, let us say, p_1 new equations. If $p+p_1 = N$, the system (2.3.3), and therefore the system (2.3.1), has only the trivial integral $u = $ constant. If $p+p_1 < N$, we repeat on the new system the same process of extension. Continuing in this way, we must obtain after a finite number of steps either a system with N equations, or a system of less than N equations for which the vanishing of all the expressions (2.3.2) is a consequence of the system of equations itself. That is, we must have, for this new system

$$[X_\sigma X_\rho - X_\rho X_\sigma](u) = c_{\rho\sigma}^\tau X_\tau(u), \tag{2.3.4}$$

where the $c_{\rho\sigma}^\tau$ are functions of the x^i. A system (2.3.1) for which (2.3.4) is valid, is known as a *complete* system. We have thus shown that every linear system (2.3.1) can be extended to a complete system having precisely the same common integrals.

For any function f, the expression $X_\rho(f)$ is an invariant; hence so also are expressions of the form $X_\sigma X_\rho(f)$. Under a transformation of independent variables, the equations (2.3.4) go over into a set of similar equations, the $c_{\rho\sigma}^\tau$ transforming as a set of scalar invariants. Hence a complete system remains complete under transformations of variables.

We shall say that two systems

$$X_\rho(u) = 0, \qquad Y_\rho(u) = 0, \tag{2.3.5}$$

are *equivalent* if we can express the quantities Y_ρ as a linear sum of the X_ρ, and vice versa. That is, the two systems are equivalent if there exists a non-singular matrix of functions (a_ρ^σ) such that

$$Y_\rho(u) = a_\rho^\sigma X_\sigma(u). \tag{2.3.6}$$

Now we will show that if one of these equivalent systems, say $X_\rho = 0$, is complete, then so is the other. In fact, we have

$$[Y_\sigma Y_\rho - Y_\rho Y_\sigma](u) = a_\sigma^\lambda X_\lambda \{a_\rho^\mu X_\mu(u)\} - a_\rho^\lambda X_\lambda \{a_\sigma^\mu X_\mu(u)\}$$
$$= a_\sigma^\lambda a_\rho^\mu (X_\lambda X_\mu - X_\mu X_\lambda)(u) + a_\sigma^\lambda X_\lambda (a_\rho^\mu) X_\mu(u) - a_\rho^\lambda X_\lambda (a_\sigma^\mu) X_\mu(u), \qquad (2.3.7)$$

and the right-hand side can, by hypothesis, be expressed as a linear combination of the $\dot{X}_\mu$. Therefore, by (2.3.6), (or rather the inverse of (2.3.6)), the expression (2.3.7) can be written as a linear sum of the Y_σ. This shows that the system $Y_\rho = 0$ is complete.

Exercise. Interpret (2.3.6) geometrically in terms of the b_ρ^i.

Next we wish to show that, if we are given a complete system, we can always find an equivalent complete system for which the quantities $c_{\rho\sigma}^\tau$ appearing in (2.3.4) vanish identically. This special kind of complete system is known as a Jacobian system; for a Jacobian system we have, therefore

$$[X_\sigma X_\rho - X_\rho X_\sigma](u) \equiv 0. \qquad (2.3.8)$$

To establish the remark just made, we note that the p equations $X_\rho = 0$ are independent by hypothesis, and can therefore be solved for p of the partial derivatives of u in terms of the remaining $N-p$ partial derivatives. In fact, the system so obtained,

$$Y_\alpha(u) \equiv \frac{\partial u}{\partial x^\alpha} + b_\alpha^\beta \frac{\partial u}{\partial x^\beta} = 0, \qquad (2.3.9)$$

where $\alpha = 1,...,p$; $\beta = p+1,...,N$, is equivalent to the original system and is therefore complete. Let us now form the commutator expressions $Y_\sigma Y_\rho - Y_\rho Y_\sigma$; it is clear that the partials $\partial u/\partial x^\alpha$ ($\alpha = 1,...,p$) will not appear. But this implies that none of the quantities Y_σ can appear; in other words that the commutators are zero, so that the system is Jacobian.

We are now in a position to attack our original problem in connexion with the system $X_\rho(u) = 0$, namely, the determination of the common integrals. From § 1 we recall that if the system consists of a single equation it has $N-1$ independent integrals. We have seen that we can extend the given system to a complete system; suppose this done. The following lemma can now be proved: *Every complete system of p equations in N variables can be reduced by the integration of any one of these equations to a complete system of $p-1$ equations in $N-1$ variables.*

Suppose, for example, that we have found $N-1$ independent integrals

f^i of the equation $X_p(u) = 0$. Choose now a new system of coordinates x'^i with

$$x'^i = f^i \quad (i = 1, 2,..., N-1), \tag{2.3.10}$$

and with x'^N independent of $f^1,..., f^{N-1}$. Then the equation $X_p(u) = 0$ goes over into the simple form

$$\frac{\partial u}{\partial x'^N} = 0.$$

The remaining $p-1$ equations may be solved for $p-1$ other derivatives, and the new system written in the form (dropping primes):

$$(a) \quad X_\alpha(u) = \frac{\partial u}{\partial x^\alpha} + b^\beta_\alpha \frac{\partial u}{\partial x^\beta} = 0 \quad (\alpha = 1,...,p-1; \, \beta = p+1,...,N),$$

$$(b) \quad X_p(u) = \frac{\partial u}{\partial x^N} = 0. \tag{2.3.11}$$

This system has the special form (2.3.9) and is therefore a Jacobian system. Consequently the commutator expressions of the X_p are all zero, so that in particular

$$[X_p X_\alpha - X_\alpha X_p](f) = \frac{\partial}{\partial x^N}\left(\frac{\partial f}{\partial x^\alpha} + b^\beta_\alpha \frac{\partial f}{\partial x^\beta}\right) - \left(\frac{\partial}{\partial x^\alpha} + b^\beta_\alpha \frac{\partial}{\partial x^\beta}\right)\frac{\partial f}{\partial x^N}$$

$$= \frac{\partial b^\beta_\alpha}{\partial x^N}\frac{\partial f}{\partial x^\beta} = 0 \quad\quad (\alpha = 1,...,p-1).$$

Since this holds for arbitrary functions f we must conclude that the functions b^β_α are independent of the variable x^N. Therefore the set (2.3.11(a)) of $p-1$ equations is a self-contained system in $(N-1)$ independent variables $x^1,..., x^{N-1}$. Moreover, since

$$[X_\alpha X_\beta - X_\beta X_\alpha](f) = 0 \quad\quad (\alpha, \beta = 1,...,p-1),$$

the system (2.3.11(a)) is a complete (indeed a Jacobian) system. Finally we note that the function u must be independent of x^N, as is shown by (2.3.11(b)), and must satisfy the complete system (2.3.11(a)). This proves the lemma.

It is now easy to foresee the entire result, since we can apply repeatedly the proposition just established and reduce the given complete system of p equations in N variables to a 'system' of 1 equation in $N-p+1$ variables. This single linear equation will have $N-p$ independent integrals, and we conclude that *every complete system of p equations in N variables has $N-p$ independent integrals*. The general integral of the system is an arbitrary function of these $N-p$ special integrals. It follows that to fix a particular solution of a complete system of p equations, a function of $N-p$ variables is required.

The following example, with two equations in four variables, serves to illustrate this integration process for linear homogeneous systems. Let

$$X_1(u) = x^1 \frac{\partial u}{\partial x^1} + x^2 \frac{\partial u}{\partial x^2} + x^3 \frac{\partial u}{\partial x^3} + x^4 \frac{\partial u}{\partial x^4} = 0,$$

$$X_2(u) = x^2 \frac{\partial u}{\partial x^1} - x^1 \frac{\partial u}{\partial x^2} = 0$$

be the system, the common integrals of which are to be found. It is easily verified that $[X_1 X_2 - X_2 X_1](u)$ is identically zero, so the system is complete. Now the integrals of $X_2(u) = 0$ are arbitrary functions of x^3, x^4, and of the first integral $(x^1)^2 + (x^2)^2 = c$ of the equations of the characteristics of $X_2(u) = 0$. Thus if we set $z^1 = (x^1)^2 + (x^2)^2$, $z^2 = x^2$, $z^3 = x^3$, $z^4 = x^4$, the equation $X_2(u) = 0$ becomes simply $\partial u/\partial z^2 = 0$. Hence u is a function of the three quantities z^1, z^3, z^4 only. The equation $X_1(u) = 0$ now takes the form

$$2z^1 \frac{\partial u}{\partial z^1} + z^3 \frac{\partial u}{\partial z^3} + z^4 \frac{\partial u}{\partial z^4} = 0,$$

and this equation has the general integral $u = f\left\{\frac{(z^3)^2}{z^1}, \frac{(z^4)^2}{z^1}\right\}$. In terms of the original independent variables, the general integral is any function of the combinations

$$\frac{(x^3)^2}{(x^1)^2 + (x^2)^2}, \qquad \frac{(x^4)^2}{(x^1)^2 + (x^2)^2}.$$

2.4. Adjoint systems. In the first section of this chapter we investigated the geometrical meaning of a homogeneous linear partial differential equation of the first order, and found that the equation states that a certain displacement lies in every integral surface. Now the reader will have noticed the geometrical duality which subsists between contravariant vectors (directions) and covariant vectors (surface elements). The type of equation whose geometric meaning is dual to that of the linear partial differential equation is the homogeneous total differential equation of the form

$$b_i \, dx^i = 0. \tag{2.4.1}$$

In this equation, b_i represents a given covariant vector field; that is, a field of surface elements. An integral of (2.4.1) is *defined* to be any function $u(x^i)$ such that

$$du = \frac{\partial u}{\partial x^i} dx^i = 0 \tag{2.4.2}$$

whenever dx^i satisfies (2.4.1). This implies, since the dx^i are otherwise independent, that the covariant vector $\partial u/\partial x^i$ is a multiple of b_i, and so defines the same surface element at P. These surface elements are tangent at each point to the *integral surface*

$$u(x^i) = \text{constant}. \tag{2.4.3}$$

We summarize the geometrical content of (2.4.1): *the 'unknown' displacement dx^i is tangent to the surface element given by b_i.* Furthermore, an integral surface $u = $ constant is any surface which envelops the given field of surface elements. Such a surface may not exist, however, since the field of surface elements does not necessarily possess an envelope.

Exercise 1. If (2.4.2) holds, and $b_1 \neq 0$, show that $\partial u / \partial x^i = \lambda b_i$ by choosing λ so that $\partial u / \partial x^1 = \lambda b_1$; then allowing $dx^2, ..., dx^N$ to vary arbitrarily.

Exercise 2. Show that $b_{ik} = \dfrac{\partial b_i}{\partial x^k} - \dfrac{\partial b_k}{\partial x^i}$

is a covariant tensor of the second order, with $b_{ik} = b_{ki}$. Show that if b_{ik} vanishes, (2.4.1) is completely integrable in the sense of Chapter I.

Exercise 3. Show that if the differential $b_i \, dx^i$ has an integrating factor μ, then

$$\frac{\partial \mu}{\partial x^k} b_i - \frac{\partial \mu}{\partial x^i} b_k + \mu b_{ik} = 0,$$

and that if the surface elements b_i have an envelope, there exists an integrating factor μ.

The geometrical significance of a system of total differential equations is now easily found. Let there be given q independent equations

$$b_i^\rho \, dx^i = 0. \tag{2.4.4}$$

(Independent means that the vectors b_i^ρ ($\rho = 1, ..., q$) are to be linearly independent, so that there is no relation

$$a_\rho b_i^\rho = 0 \qquad (i = 1, ..., N),$$

where the a_ρ are functions of position. In other words, the rank of the $N \times q$ matrix of components b_i^ρ is to have its full value q.) Now the q conditions (2.4.4) restrict the displacement dx^i to a family of $p = N - q$ dimensions. Analytically, it follows from the independence of the b_i^ρ and from the theory of linear equations, that there exist $p = N - q$ orthogonal contravariant vectors b_σ^i ($\sigma = 1, ..., p$) satisfying

$$b_i^\rho b_\sigma^i = 0. \tag{2.4.5}$$

We may certainly assume that the b_σ^i transform as contravariant vectors defined in each, or any, coordinate system by (2.4.5). It is also clear that the set of p vectors b_σ^i can be chosen in many ways, but that the p-dimensional direction spanned by them is determined by the $q = (N-p)$-dimensional family of surface elements b_i^ρ.

Let us define a common integral of (2.4.4) to be any function $u(x^i)$ such that

$$du = \frac{\partial u}{\partial x^i} dx^i = 0 \tag{2.4.6}$$

whenever *all* of (2.4.4) hold. It is now evident that (2.4.4) holds if and only if the displacement dx^i is a multiple sum of the vectors b_σ^i:

$$dx^i = b_\sigma^i \, d\theta^\sigma. \tag{2.4.7}$$

Here the $d\theta^\sigma$ are suitable infinitesimals which can be varied independently of one another. Then from (2.4.6) and (2.4.7) we have

$$du = \frac{\partial u}{\partial x^i} dx^i = \frac{\partial u}{\partial x^i} b_\sigma^i \, d\theta^\sigma = 0,$$

whence

$$b_\sigma^i \frac{\partial u}{\partial x^i} = 0 \tag{2.4.8}$$

for $\sigma = 1,..., p = N-q$. Therefore an integral of the q total differential equations (2.4.4) satisfies the $p = N-q$ linear partial differential equations (2.4.8). Conversely, it is also evident that any common integral of the system (2.4.8) is an integral of the total differential equations (2.4.4). The system (2.4.4) of total differential equations and the system (2.4.8) of partial differential equations are said to be dual, or adjoint, to one another.

Exercise. Any system, equivalent to (2.4.8) in the sense of § 2.3, may appear in place of (2.4.8) due to a different choice of the b_σ^i; and only systems equivalent to (2.4.8) may so appear.

In particular, let us examine the set of $N-1$ total differential equations which is adjoint to a single linear homogeneous partial differential equation

$$b^i \frac{\partial u}{\partial x^i} = 0. \tag{2.4.9}$$

Now there exist $N-1$ independent covariant vectors b_i^ρ such that

$$b_i^\rho b^i = 0, \tag{2.4.10}$$

for $\rho = 1,..., N-1$. The adjoint system is

$$b_i^\rho \, dx^i = 0. \tag{2.4.11}$$

Since the rank of the matrix of components b_i^ρ is $N-1$, equations

(2.4.11) determine the ratios of the dx^i uniquely. In view of (2.4.10), therefore, we have

$$dx^i = b^i \, dt \qquad (2.4.12)$$

for some infinitesimal dt. These are just the equations of the characteristics of (2.4.9). We have seen that the integral surfaces of (2.4.9) are precisely those surfaces which contain the directions of the characteristic curves at each point.

Returning to the general case of the system of partial differential equations (2.4.8), we see that the adjoint system (2.4.4) consisting of $q = N - p$ equations has, not one-dimensional characteristic directions, but p-dimensional characteristic surface elements. The integral surfaces of the system (2.4.8) are precisely those surfaces which at each point contain these p-dimensional characteristic surface elements.

In the preceding section we saw that the common integrals of a system of linear equations like (2.4.8) all satisfy the equations of a complete system which may include some additional equations. Thus the adjoint system of total differential equations may be reduced in number corresponding to this increase in the number of partial differential equations. In actual fact we may assume that the system (2.4.8) has been extended to a complete system and written in the Jacobian form

$$X_\alpha(u) = \frac{\partial u}{\partial x^\alpha} + b_\alpha^\rho \frac{\partial u}{\partial x^\rho} = 0, \qquad (2.4.13)$$

where $\alpha = 1, ..., p$; $\rho = p+1, ..., N$. The adjoint system of $(N-p)$ total differential equations corresponding to (2.4.13) can now be written explicitly in an equally simple form, namely,

$$dx^\rho = b_\alpha^\rho \, dx^\alpha. \qquad (2.4.14)$$

We shall leave to the reader the verification of the relations corresponding to (2.4.5) in this case. Note that in the $q = N - p$ equations of (2.4.14) the last q variables x^ρ have been transposed to the left-hand side, where they may be regarded as dependent variables.

The integral surface elements of (2.4.14) are the characteristic surface elements of the Jacobian system (2.4.13), and they are of dimension p. By an integral surface element of dimension p is meant an element of a p-dimensional surface such that $q = N - p$ relations—in this case (2.4.14)—hold for the displacements on the surface element. In the case $p = 1$, we saw that the one-dimensional characteristic surface elements, namely, the characteristic directions, give rise upon integration to characteristic curves; or, if it is preferred, characteristic one-dimensional 'surfaces'. Furthermore these curves each lay entirely in a single integral surface of the partial differential equation. Now, in the present case of p equations, we would like to show that the p-dimensional characteristic surface elements defined by (2.4.14) can be integrated (pieced smoothly together) into p-dimensional characteristic surfaces.

This will be accomplished if we can show that (2.4.14) has solutions

$$x^\rho = x^\rho(x^\alpha), \qquad (2.4.15)$$

for $\rho = p+1,...,N$; $\alpha = 1,...,p$. The equations (2.4.15) are then the equations of the p-dimensional characteristic strips of the Jacobian system (2.4.13). We therefore require an existence theorem of the type of § 1.3. For analytic systems of equations like (2.4.14), such an existence theorem can be proved in essentially the same way as in the case of a single equation. We remark that the conditions of integrability now take the form

$$\frac{\partial b_\alpha^\rho}{\partial x^\beta} + b_\beta^\sigma \frac{\partial b_\alpha^\rho}{\partial x^\sigma} = \frac{\partial b_\beta^\rho}{\partial x^\alpha} + b_\alpha^\sigma \frac{\partial b_\beta^\rho}{\partial x^\sigma}. \qquad (2.4.16)$$

Here $\alpha, \beta = 1,...,p$; $\rho, \sigma = p+1,...,N$, there being one set of conditions for each value of the index ρ. In order to apply our proof, we must show that these conditions are identically satisfied. To do this, we need only use the fact that the system (2.4.13) is Jacobian; that is, the expressions $[X_\alpha X_\beta - X_\beta X_\alpha](u)$ vanish identically. We know that these expressions contain no second derivatives of f in any case, so that in our computation they may be omitted. We have

$$X_\alpha X_\beta(f) = \left(\frac{\partial}{\partial x^\alpha} + b_\alpha^\rho \frac{\partial}{\partial x^\rho}\right)\left(\frac{\partial f}{\partial x^\beta} + b_\beta^\sigma \frac{\partial f}{\partial x^\sigma}\right)$$

$$= \frac{\partial b_\beta^\sigma}{\partial x^\alpha} \frac{\partial f}{\partial x^\sigma} + b_\alpha^\rho \frac{\partial b_\beta^\sigma}{\partial x^\rho} \frac{\partial f}{\partial x^\sigma} + \cdots. \qquad (2.4.17)$$

Since the terms with second derivatives all cancel out, we find

$$[X_\alpha X_\beta - X_\beta X_\alpha](f) = \left[\frac{\partial b_\beta^\rho}{\partial x^\alpha} + b_\alpha^\sigma \frac{\partial b_\beta^\rho}{\partial x^\sigma} - \frac{\partial b_\alpha^\rho}{\partial x^\beta} - b_\beta^\sigma \frac{\partial b_\alpha^\rho}{\partial x^\sigma}\right]\frac{\partial f}{\partial x^\rho}. \qquad (2.4.18)$$

Since the right-hand side must vanish identically, and $\partial u/\partial x^\rho$ is not zero in general, it follows that the expression in the square bracket vanishes, and therefore that the integrability conditions (2.4.16) hold. The converse of the formal result just proved can be shown by reversing the argument. *Thus the adjoint system of total differential equations is completely integrable if and only if the given system of partial differential equations is Jacobian.*

Consequently, if we are given a complete system (2.4.13) consisting of p independent equations $X_\alpha(u) = 0$, the adjoint system of total differential equations, which is completely integrable, has solutions of the form of

(2.4.15). The surfaces of p dimensions represented by these equations are the p-dimensional characteristic strips of the given complete system; that is, every integral surface of the given system contains at each point the p-dimensional surface element tangent to the characteristic strip. Again, as in the case of a single equation, the integral surfaces themselves may be thought of as constructed from families of the characteristic strips.

As an illustration of these relationships, let us take the following system of two partial differential equations in three variables:

$$\frac{\partial u}{\partial x^1} + x^2 \frac{\partial u}{\partial x^3} = 0, \qquad \frac{\partial u}{\partial x^2} + x^1 \frac{\partial u}{\partial x^3} = 0.$$

It is easy to show directly that this system is complete, and to deduce the form of the general integral, which will be a function of one variable, by the method of the preceding section. However, it is seen that the adjoint system consists of the single equation

$$dx^3 = x^2 \, dx^1 + x^1 \, dx^2,$$

which has the integral $x^3 - x^1 x^2 = $ constant. The equation $x^3 - x^1 x^2 = $ constant represents a two-dimensional characteristic 'strip'. The general integral of the system of partial differential equations must be of the form

$$u = f(x^3 - x^1 x^2),$$

and it is easily verified that this is correct. An integral surface $u = $ constant must in this case consist of a single characteristic two-dimensional surface in the three-dimensional space of the independent variables.

Exercise 1. Show that the system

$$x^1 \frac{\partial u}{\partial x^1} - x^2 \frac{\partial u}{\partial x^2} + x^2 \frac{\partial u}{\partial x^3} = 0, \qquad x^1 \frac{\partial u}{\partial x^1} + (x^2 + x^3) \frac{\partial u}{\partial x^3} = 0$$

is complete, and find a general integral of the system.

Exercise 2. Solve the quasi-linear equations:

(a) $\quad x^1 \dfrac{\partial z}{\partial x^1} + x^2 \dfrac{\partial z}{\partial x^2} = -\dfrac{2(x^1 z - x^2 z + x^1 x^2)}{4x^2 - x^1 + z};$

(b) $\quad z \dfrac{\partial z}{\partial x^1} - z \dfrac{\partial z}{\partial x^2} = z^2 + (x^1 + x^2)^2.$

Exercise 3. If a system of linear homogeneous first-order equations has an integral of the form $u = \sum\limits_{\alpha=1}^{p} a_\alpha f_\alpha(x^i)$, for arbitrary values of the a_α, then the more general form

$$u = F(f_1,...,f_p),$$

for arbitrary functions F, is also an integral.

Exercise 4. Euler's theorem on homogeneous functions: the equation

$$x^i \frac{\partial u}{\partial x^i} = au,$$

has as general integral any function homogeneous of degree a in the x^i.

Exercise 5. If $z = h(x^i)$ is a particular solution of the non-homogeneous linear equation

$$b^i(x^j) \frac{\partial z}{\partial x^i} = b(x^j),$$

show that every solution has the form $z = h(x^i) + \pi\{f_{(\alpha)}(x^j)\}$, where the f_α are $N-1$ independent solutions of the abbreviated equation. Reduce the linear equation $b^i \, \partial z/\partial x^i = bz$ to the above form.

Exercise 6. Given the Jacobian system of two equations in four variables with

$$b_1^i = (1, 0, x^4, x^3),$$

$$b_2^i = (0, 1, x^3, x^4),$$

write down the adjoint system, and integrate it. Also find the general integral of the given Jacobian system.

III

NON-LINEAR EQUATIONS OF THE FIRST ORDER

ANY partial differential equation of first order is a statement of condition on the first derivatives of the dependent variable. A geometrical interpretation of this statement would be: the surface element tangent at any point of an integral surface must satisfy some given geometrical condition. For the homogeneous linear equation this condition took an especially simple form—the integral surface contains a given displacement. Moreover, this interpretation led naturally to the definition of the characteristic curves, from which in turn the solution of the initial value problem was found. We must now extend these concepts and methods to the general case where the geometric condition upon the tangent surface elements may take an arbitrary form. The method of Cauchy, which we follow, leads here also to a solution of the initial value problem.

A quite different approach to the non-linear equation can be made via the theory of the complete integral developed by Lagrange, and later Jacobi. A complete integral is a certain general form of solution involving a number of independent parameters, such that any solution of the equation can be found from it by differentiation and elimination processes. The study of complete integrals is important in analytical mechanics, where the characteristic curves may be interpreted as the paths or motions of a dynamical system; these may be found if a complete integral is known. Lastly in this chapter we consider systems of non-linear equations of first order. The integration theory for these leads to an analytical method of determining a complete integral.

3.1. Geometric theory of the characteristics. The most general form for a single partial differential equation with one dependent variable u is

$$F(x^i, u, p_i) = 0, \qquad (3.1.1)$$

where the notation

$$p_i = \frac{\partial u}{\partial x^i} \qquad (3.1.2)$$

has been used. We see that p_i is a covariant vector under transformation of the N independent variables x^i. An integral surface S of (3.1.1) is a surface of N dimensions $u = f(x^i)$ in the space V_{N+1} of the variables u and x^i ($i = 1,..., N$), where $f(x^i)$ is a solution function of (3.1.1). This differential equation has the form of a condition imposed on the surface element, represented by p_i, which is tangent to the integral surface. The nature of the function F determines how this condition varies from point to point in V_{N+1} (Fig. 3).

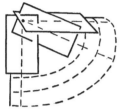

FIG. 3. Integral surface elements of $F(x^i, u, p_i) = 0$.

The essentially geometric method of solution for this equation, which was discovered by Cauchy, is a generalization of the method of characteristic curves just described for the linear equation. Now the characteristic direction for the linear equation was given by the contravariant vector b^i. Now, for the non-linear equation, we seek an analogous contravariant vector. Such a vector is provided by certain partial derivatives of the function F. Indeed, at a fixed point P of coordinates x^i, we have

$$p_k = \frac{\partial x'^i}{\partial x^k} p_i'. \tag{3.1.3}$$

If, in (3.1.3), we regard the p_k for the moment as functions of the p_i' alone, we see that

$$\frac{\partial p_k}{\partial p_i'} = \frac{\partial x'^i}{\partial x^k}. \tag{3.1.4}$$

Therefore,

$$\frac{\partial F}{\partial p_i'} = \frac{\partial F}{\partial p_k} \frac{\partial p_k}{\partial p_i'} = \frac{\partial F}{\partial p_k} \frac{\partial x'^i}{\partial x^k}, \tag{3.1.5}$$

and this equation states that $\partial F/\partial p_k$ is indeed a contravariant vector. This vector, which in general depends upon all the variables x^i, u, and p_i, determines a certain infinitesimal displacement, the 'characteristic' displacement

$$dx^i = \frac{\partial F}{\partial p_i} dt. \tag{3.1.6}$$

On an integral surface, we have $u = u(x^i)$, and so

$$du = p_i \, dx^i = p_i \frac{\partial F}{\partial p_i} dt. \tag{3.1.7}$$

Note that the right-hand side of (3.1.7) is an invariant.

Exercise 1. What is $\partial F/\partial p_i$ in the case of a linear equation? When is the characteristic displacement dependent only on the x^i and u?

Exercise 2. Show that the displacement (3.1.6)–(3.1.7) in V_{N+1} lies on the envelope of the surface elements at P which satisfy the differential equation.

On an integral surface $u = u(x^i)$ the derivatives p_i are determinate functions of the x^i. Let us find equations which will describe the change of the p_i under the characteristic displacement (3.1.6). By hypothesis, (3.1.1) is satisfied, and on being differentiated with respect to x^i it yields

$$\frac{\partial F}{\partial x^k} + \frac{\partial F}{\partial u} p_k + \frac{\partial F}{\partial p_i} \frac{\partial p_i}{\partial x^k} = 0.$$

Since

$$\frac{\partial p_i}{\partial x^k} = \frac{\partial^2 u}{\partial x^i \partial x^k} = \frac{\partial p_k}{\partial x^i}, \tag{3.1.8}$$

we see that

$$dp_k = \frac{\partial p_k}{\partial x^i} dx^i = \frac{\partial p_i}{\partial x^k} \frac{\partial F}{\partial p_i} dt \tag{3.1.9}$$

$$= -\left(\frac{\partial F}{\partial x^k} + p_k \frac{\partial F}{\partial u}\right) dt,$$

from (3.1.6), (3.1.8), and (3.1.9). The equations (3.1.6), (3.1.7), and (3.1.9) together form a self-contained system of $2N+1$ ordinary differential equations which we may write together in the form

$$\frac{dx^i}{dt} = \frac{\partial F}{\partial p_i}; \quad \frac{du}{dt} = p_i \frac{\partial F}{\partial p_i}; \quad \frac{dp_i}{dt} + \frac{\partial F}{\partial x_i} + p_i \frac{\partial F}{\partial u} = 0. \tag{3.1.10}$$

These are the equations of the characteristic strips, and are the generalization of (2.2.5) to the non-linear equation.

It is interesting to verify that (3.1.10) is covariant under transformations of the coordinate variables x^i. The first group of N equations is certainly covariant, since each side is a contravariant vector; the single equation for u is also, since it contains only the invariant $p_i(\partial F/\partial p_i)$. In the third group of equations, the third term is a covariant vector. We will see that either of the two first terms separately is not a vector, but that the sum of the two is. Differentiate (3.1.3) with respect to t:

$$\frac{dp'_k}{dt} = \frac{\partial x^i}{\partial x'^k} \frac{dp_i}{dt} + p_i \frac{\partial^2 x^i}{\partial x'^k \partial x'^m} \frac{dx'^m}{dt}. \tag{3.1.11}$$

Now from first principles,

$$F\left(x'^k, u(x'^k), \frac{\partial x^i}{\partial x'^m} p_i\right) = F(x^i(x'^k), u(x'^k), p_i), \tag{3.1.12}$$

and if we now differentiate with respect to x'^k, there results

$$\frac{\partial F}{\partial x'^k} + \frac{\partial F}{\partial p'_m} \frac{\partial^2 x^i}{\partial x'^k \partial x'^m} p_i = \frac{\partial F}{\partial x^i} \frac{\partial x^i}{\partial x'^k}. \tag{3.1.13}$$

Now add (3.1.11) and (3.1.13). We find

$$\frac{dp'_k}{dt}+\frac{\partial F}{\partial x'^k} = \frac{\partial x^i}{\partial x'^k}\left(\frac{dp_i}{dt}+\frac{\partial F}{\partial x^i}\right)+p_i\frac{\partial^2 x^i}{\partial x'^k\partial x'^m}\left(\frac{dx'^m}{dt}-\frac{\partial F}{\partial p'_m}\right). \quad (3.1.14)$$

From the first of (3.1.10), imagined written in the x'^i coordinates, the term containing the second derivative vanishes, and (3.1.14) is then a statement that the two terms in question transform as a covariant vector. If, therefore, we are given a solution curve. namely x^i, u, and p_i as functions of t satisfying (3.1.10), the differential equations (3.1.10) are tensor equations, that is, they are covariant.

Geometrically, a solution

$$x^i = x^i(t), \qquad u = u(t), \qquad p_i = p_i(t) \qquad (3.1.15)$$

of (3.1.10) represents a characteristic strip; that is, a curve having a surface element defined at each point. The relation

$$\frac{du}{dt} = p_i\frac{dx^i}{dt}, \qquad (3.1.16)$$

which follows from (3.1.10), tells us that at each point the curve is tangent to the surface element at that point. Relative to the differential equation, the essential property of the strip (3.1.15) is that, if one surface element (x^i, u, p_i) satisfies the equation, then so do all other surface elements of the strip. To prove this, we first show that the relation

$$F(x^i, u, p_i) = \text{constant} \qquad (3.1.17)$$

is a first integral of (3.1.10). This follows since, along the strip, we have

$$\frac{dF}{dt} = \frac{\partial F}{\partial x^i}\frac{dx^i}{dt}+\frac{\partial F}{\partial u}\frac{du}{dt}+\frac{\partial F}{\partial p_i}\frac{dp_i}{dt}$$

$$= \frac{\partial F}{\partial x^i}\frac{\partial F}{\partial p_i}+\frac{\partial F}{\partial u}\,p_i\frac{\partial F}{\partial p_i}-\frac{\partial F}{\partial p_i}\left(\frac{\partial F}{\partial x^i}+p_i\frac{\partial F}{\partial u}\right) = 0. \qquad (3.1.18)$$

If now $F(x^i, u, p_i) = 0$ for $t = t_0$, the same holds for all values of t, and so each surface element of the strip satisfies the differential equation.

As a corollary to this result, it follows that if two integral surfaces have a common surface element, that is, if they are tangent at a single point, then they contain a common characteristic strip and so are tangent at all points of a curve. It is to be noted that this does not necessarily hold for two integral surfaces which intersect without being tangent. These results should be compared with the corresponding statements for the linear equation.

We shall now turn to the problem of constructing solutions of the partial differential equation by means of the characteristic strips. The appropriate initial value problem is stated in the same way as for the linear equation: given a V_{N-1} in V_{N+1} find an integral surface $S \equiv V_N : u = u(x^i)$ which contains V_{N-1}. Suppose that V_{N-1} is given parametrically as follows:

$$x^i = x^i(t^1,...,t^{N-1}), \qquad u = u(t^1,...,t^{N-1}). \qquad (3.1.19)$$

We must determine at each point of V_{N-1} a surface element in V_N which is tangent to V_{N-1} and also satisfies the partial differential equation. The

FIG. 4. Integral surface generated by characteristic strips.

quantities p_i to be determined as functions of the t^α ($\alpha = 1,..., N-1$) must therefore satisfy the $N-1$ conditions

$$\frac{\partial u}{\partial t^\alpha} = p_i \frac{\partial x^i}{\partial t^\alpha}, \qquad (3.1.20)$$

as well as the original equation $F(x^i, u, p_i) = 0$. We shall suppose that these N conditions enable us to determine the $p_i = p_i(t^\alpha)$. This will indeed be the case if, as we shall now assume, the determinant

$$\Delta = \left| \frac{\partial F}{\partial p_i}, \frac{\partial x^i}{\partial t^1},..., \frac{\partial x^i}{\partial t^{N-1}} \right| \equiv \left| \frac{\partial F}{\partial p_i}, \frac{\partial x^i}{\partial t^\alpha} \right| \neq 0.$$

Let $\qquad x^i = x^i(t, t^\alpha), \qquad u = u(t, t^\alpha), \qquad p_i = p_i(t, t^\alpha), \qquad (3.1.21)$

be the solutions of the characteristic equations with these initial values. The existence and uniqueness of these functions for some range of values of t will be assumed from the theory of ordinary differential equations. In view of (3.1.1) and (3.1.18), we see that (3.1.21) is an N-parameter family of surface elements, each of which satisfies the partial differential equation. The problem before us is to show that these surface elements can be pieced together smoothly into a common surface, which will then be an integral surface containing V_{N-1}, as required (Fig. 4). From (3.1.10) we see that

$$\frac{\partial u}{\partial t} = p_i \frac{\partial x^i}{\partial t}, \qquad (3.1.22)$$

that is, the surface elements are tangent to the characteristic curves. We must show that similar relations

$$U_\alpha \equiv \frac{\partial u}{\partial t^\alpha} - p_i \frac{\partial x^i}{\partial t^\alpha} = 0 \qquad (3.1.23)$$

hold for each of the parameters t^α. According to (3.1.20) these relations hold for $t = 0$, i.e. on V_{N-1}. Now

$$\frac{dU_\alpha}{dt} = \frac{d}{dt}\left(\frac{\partial u}{\partial t^\alpha}\right) - \frac{dp_i}{dt}\frac{\partial x^i}{\partial t^\alpha} - p_i\frac{d}{dt}\left(\frac{\partial x^i}{\partial t^\alpha}\right)$$

$$= \frac{\partial}{\partial t^\alpha}\left(\frac{du}{dt}\right) - \frac{dp_i}{dt}\frac{\partial x^i}{\partial t^\alpha} - p_i\frac{\partial}{\partial t^\alpha}\left(\frac{dx^i}{dt}\right). \qquad (3.1.24)$$

Differentiating (3.1.22) with respect to t^α, we see that

$$\frac{d}{dt}\left(\frac{\partial u}{\partial t^\alpha}\right) = \frac{\partial}{\partial t^\alpha}\left(\frac{du}{dt}\right) = \frac{\partial p_i}{\partial t^\alpha}\frac{dx^i}{dt} + p_i\frac{\partial}{\partial t^\alpha}\left(\frac{dx^i}{dt}\right), \qquad (3.1.25)$$

and, adding this to (3.1.24), we find

$$\frac{dU_\alpha}{dt} = \frac{\partial p_i}{\partial t^\alpha}\frac{dx^i}{dt} - \frac{dp_i}{dt}\frac{\partial x^i}{\partial t^\alpha} = \frac{\partial F}{\partial p_i}\frac{\partial p_i}{\partial t^\alpha} - \left(\frac{\partial F}{\partial x^i} + p_i\frac{\partial F}{\partial u}\right)\frac{\partial x^i}{\partial t^\alpha}. \qquad (3.1.26)$$

But
$$\frac{\partial F}{\partial t^\alpha} = \frac{\partial F}{\partial x^i}\frac{\partial x^i}{\partial t^\alpha} + \frac{\partial F}{\partial u}\frac{\partial u}{\partial t^\alpha} + \frac{\partial F}{\partial p_i}\frac{\partial p_i}{\partial t^\alpha} = 0, \qquad (3.1.27)$$

since every surface element satisfies the differential equation. Subtracting (3.1.27) from (3.1.26) we find that several terms cancel, and so obtain

$$\frac{dU_\alpha}{dt} = -\frac{\partial F}{\partial u}\left(\frac{\partial u}{\partial t^\alpha} - p_i\frac{\partial x^i}{\partial t^\alpha}\right) = -\frac{\partial F}{\partial u}U_\alpha. \qquad (3.1.28)$$

This ordinary differential equation integrates to

$$U_\alpha(t) = U_\alpha(0)\exp\left[-\int_0^t \frac{\partial F}{\partial u}\,d\tau\right]. \qquad (3.1.29)$$

Since $U_\alpha(0) = 0$, we conclude that U_α vanishes identically. Therefore (3.1.23) is proved and the surface elements (3.1.21) are contained in a common surface, which is an integral surface, and satisfies the conditions of the initial value problem.

In order to construct the solution function u as a function of the x^i, we must solve the first N equations (3.1.21) for t and the t^α as functions of the x^i, and then substitute the expressions so formed in $u(t, t^\alpha)$. This process is possible if the functional determinant

$$\Delta(t) = \frac{\partial(x^1, \ldots, x^N)}{\partial(t, t^1, \ldots, t^{N-1})} \equiv \left| \frac{\partial x^i}{\partial t}, \frac{\partial x^i}{\partial t^\alpha} \right| \neq 0. \tag{3.1.30}$$

In view of (3.1.10), this determinant has for $t = 0$ the value of the determinant Δ which by hypothesis differs from zero. Since (3.1.30) is a continuous function of t, it is non-zero in some t-interval about $t = 0$, and in this interval the solution function u can be expressed as a well-determined function of the independent variables.

Exercise. Interpret geometrically the failure to express u as $u(x^i)$.

We consider briefly the exceptional case relative to the initial value problem just treated, namely, when on the initial manifold V_{N-1}

$$\Delta \equiv \left| \frac{\partial F}{\partial p_i}, \frac{\partial x^i}{\partial t^\alpha} \right| = 0. \tag{3.1.31}$$

For simplicity, suppose Δ is identically zero. As in the case of the quasilinear equation, it turns out that further conditions are necessary in order that the initial value problem should be solvable, but that if these conditions are satisfied, an infinity of solutions exist.

From the vanishing of Δ it follows that there exists a proportionality of the form

$$\frac{\partial F}{\partial p_i} = \sum_\alpha \lambda^\alpha \frac{\partial x^i}{\partial t^\alpha}. \tag{3.1.32}$$

The initial manifold parametrized by the t^α is $(N-1)$-dimensional. Suppose that there exists an integral surface containing this initial manifold. That is, suppose there exists a function $u(x^i)$, with $F(x^i, u, p_i) \equiv 0$ on the initial manifold in particular. Then

$$p_i \frac{\partial F}{\partial p_i} = \sum_\alpha \lambda^\alpha \frac{\partial x^i}{\partial t^\alpha} \frac{\partial u}{\partial x^i} = \sum_\alpha \lambda^\alpha \frac{\partial u}{\partial t^\alpha}. \tag{3.1.33}$$

Also, equations of the form (3.1.7) and (3.1.8) are valid. Thus

$$-\frac{\partial F}{\partial x^k} - p_k \frac{\partial F}{\partial u} = \sum_\alpha \lambda^\alpha \frac{\partial x^i}{\partial t^\alpha} \frac{\partial p_k}{\partial x^i} = \sum_\alpha \lambda^\alpha \frac{\partial p_k}{\partial t^\alpha}. \tag{3.1.34}$$

We will show that an initial manifold of $N-1$ dimensions with surface elements p_i associated with each point, and which satisfies the three sets of conditions (3.1.32), (3.1.33), and (3.1.34), is in fact built up from an $(N-1)$-

parameter family of characteristic strips. For this reason we shall call it a characteristic strip manifold.

To show that the initial manifold is constructed of characteristic strips, we must trace upon it the characteristic curves which carry the surface elements of the strips. These curves are defined by the following differential equations among the parameters t^α of the initial manifold:

$$\frac{dt^\alpha}{ds} = \lambda^\alpha(t^\beta). \tag{3.1.35}$$

The integral curves of (3.1.35) form an $(N-2)$-parameter family which covers the initial manifold. On each of these curves we have $t^\alpha = t^\alpha(s)$, hence

$$x^i = x^i(t^\alpha) = x^i(s), \qquad u = u(t^\alpha) = u(s),$$
$$p_i = p_i(t^\alpha) = p_i(s), \tag{3.1.36}$$

by substitution. These functions of s constitute a characteristic strip, for we have from (3.1.32)–(3.1.36),

$$\frac{dx^i}{ds} = \sum_\alpha \frac{\partial x^i}{\partial t^\alpha} \frac{dt^\alpha}{ds} = \sum_\alpha \lambda^\alpha \frac{\partial x^i}{\partial t^\alpha} = \frac{\partial F}{\partial p_i},$$
$$\frac{du}{ds} = \sum_\alpha \frac{\partial u}{\partial t^\alpha} \frac{dt^\alpha}{ds} = \sum_\alpha \lambda^\alpha \frac{\partial u}{\partial t^\alpha} = p_i \frac{\partial F}{\partial p_i}, \tag{3.1.37}$$
$$\frac{dp_i}{ds} = \sum_\alpha \frac{\partial p_i}{\partial t^\alpha} \frac{dt^\alpha}{ds} = \sum_\alpha \lambda^\alpha \frac{\partial p_i}{\partial t^\alpha} = -\frac{\partial F}{\partial x^i} - p_i \frac{\partial F}{\partial u}.$$

Equations (3.1.37) are the equations of a characteristic strip. Therefore the initial manifold is generated by an $(N-1)$-parameter family of characteristic strips, and is therefore a characteristic strip manifold. We have thus shown that if $\Delta = 0$ on the initial manifold, a necessary condition that a solution exist is that the initial manifold be a characteristic strip manifold. The above reasoning can be reversed, and it follows that any surface of $N-1$ dimensions generated by characteristic strips satisfies conditions (3.1.32)–(3.1.34).

Finally we shall indicate that a solution of the initial value problem does exist if the initial manifold is a characteristic strip manifold. On the initial manifold choose an $N-2$ dimensional manifold M which has one point in common with each characteristic curve (3.1.37). This manifold M can be extended to a manifold V_{N-1} of $N-1$ dimensions which intersects the characteristic strip manifold in M, and for which $\Delta \neq 0$. Now we can solve the initial value problem for V_{N-1}; the resulting integral surface contains all characteristic strips passing through M, and therefore contains the given characteristic strip manifold. Thus a solution of the problem has been found. In fact, we can choose V_{N-1} in infinitely many ways which will lead to infinitely many solutions.

To sum up: *If $\Delta \neq 0$ on the initial manifold, the solution exists and is unique. If $\Delta \equiv 0$ on the initial manifold, then, in order that a solution exist, the initial manifold must be a characteristic strip manifold. This condition is also sufficient, and infinitely many solutions exist if it is satisfied.*

Exercise 1. Write down, and integrate, the equations of the characteristic strips for

(a) $p_1 p_2 = u$,

(b) $x^1(p_1)^2 + x^2(p_2)^2 = 0$,

(c) $p_1^2 + p_2^2 + \ldots + p_N^2 = 1$.

Exercise 2. Show that the characteristic strips of $p_1^2 = p_2$, which pass through the origin ($u = 0$, $x^1 = x^2 = 0$) generate the cone $(x^1)^2 = 4x^2 u$.

Exercise 3. If $F(x^i, u, p_i)$ is a homogeneous function of the variables p_i, then the characteristic strips of $F = 0$ lie in hyperplanes $u = $ constant.

Exercise 4. Find the solution function $u(x^1, x^2)$ of $p_1^2 + p_2^2 = 1$, which has the value u_0 on the line $ax^1 + bx^2 = 0$.

3.2. Complete integrals. The theory of the characteristic strips and the initial value problem have been presented in the form suitable to the most general partial differential equation of the first order. There will result an economy of space if we make a formal reduction of the most general equation to an equation in which the dependent variable does not appear explicitly. The idea behind this reduction has already been used in connexion with the quasi-linear equation in Chapter II. Let

$$F(x^i, u, p_i) = 0 \tag{3.2.1}$$

be the given equation. Let us define solutions $u(x^i)$ implicitly by means of a relation
$$z(u, x^i) = c,$$

where c is a constant. Thus

$$\frac{\partial z}{\partial x^i} + p_i \frac{\partial z}{\partial u} = 0,$$

and we see that z satisfies an equation of the form

$$F(x^i, p_i) = 0 \tag{3.2.2}$$

in which the number of independent variables has increased by one. Thus the space of the independent variables of the new equation is the augmented space of the independent and dependent variables of the original equation.

Consider, then, an equation of the form (3.2.2) in N independent variables x^i, and let the dependent variable be denoted by u. If u is a solution of (3.2.2), then so is $u+c$, where c is an additive constant. At any point P in the space of the variables x^i, the components p_i of the surface element tangent to any integral surface must satisfy the single condition (3.2.2). Thus the integral surface element may be thought of as belonging to an $(N-1)$-parameter family of such surface elements, all of which satisfy (3.2.2). Now let

$$u = \varphi(x^j, a_\rho) + c \tag{3.2.3}$$

be a solution of the partial differential equation which contains $N-1$ disposable constants a_ρ. Thus (3.2.3) in general defines an N-parameter family of surfaces

$$\varphi(x^j, a_\rho) + c = 0$$

in the N-dimensional space. For each value of the constants a_ρ let c be chosen so that the surface passes through P. Thus an $(N-1)$-fold infinity of these integral surfaces contain P. Their tangent surface elements at P will form an $(N-1)$-parameter family of integral surface elements if the components p_i depend independently on each of the numbers a_ρ, that is, if the $N \times (N-1)$ matrix

$$\left(\frac{\partial p_i}{\partial a_\rho}\right) = \left(\frac{\partial^2 \varphi}{\partial x^i \partial a_\rho}\right) \tag{3.2.4}$$

is of rank $N-1$. Suppose that this condition is satisfied. Then (3.2.3) is said to be a *complete integral* of the partial differential equation (3.2.2). We see that, at any point P, there is a set of values of the numbers a_ρ such that the surface element tangent to the integral surface $\varphi(x^i, a_\rho) + c = 0$ is any given integral surface element satisfying (3.2.2).

The important property of a complete integral is that *from it we can, by differentiations and eliminations only, derive the equation of any integral surface whatever.* Let S be an integral surface; the surface element at any point P tangent to S is an integral surface element; and to it there correspond values of the parameters a_ρ, which values will of course depend upon P. Thus, functions

$$a_\rho = a_\rho(x^i) \tag{3.2.5}$$

are defined by this integral surface, so that, on the surface, we have

$$u = \varphi\{x^i, a_\rho(x^i)\} + c, \tag{3.2.6}$$

and

$$p_i \equiv \frac{\partial u}{\partial x^i} = \frac{\partial}{\partial x^i} \varphi\{x^i, a_\rho(x^i)\}. \tag{3.2.7}$$

The equation (3.2.2) is satisfied by the p_i, as defined by (3.2.7), no matter how the a_ρ depend upon the x^i, simply because (3.2.3) is an integral for all values of the a_ρ. Now if we differentiate (3.2.6) with respect to x^i, and allow for the dependence of the $a_\rho(x^i)$, we see that

$$\sum_\rho \frac{\partial \varphi}{\partial a_\rho} \frac{\partial a_\rho}{\partial x^i} = 0 \qquad (3.2.8)$$

in view of (3.2.7). From the homogeneous linear algebraic equations (3.2.8) it follows that the rank of the matrix

$$\left(\frac{\partial a_\rho}{\partial x^i} \right) \qquad (3.2.9)$$

is at most $N-2$, and therefore that there exists at least one relation among the a_ρ, which is quite independent of the x^i. Let us suppose that there is exactly one such relation, say

$$a_{N-1} = \psi(a_1, ..., a_{N-2}); \qquad (3.2.10)$$

in this case the rank of (3.2.9) is equal to $N-2$. In view of (3.2.10) we see that (3.2.8) becomes

$$\sum_{\rho=1}^{N-2} \frac{\partial a_\rho}{\partial x^i} \left(\frac{\partial \varphi}{\partial a_\rho} + \frac{\partial \varphi}{\partial a_{N-1}} \frac{\partial \psi}{\partial a_\rho} \right) = 0. \qquad (3.2.11)$$

Since the rank of (3.2.9) is $N-2$, it follows from (3.2.11) that the separate equations

$$\frac{\partial \varphi}{\partial a_\rho} + \frac{\partial \varphi}{\partial a_{N-1}} \frac{\partial \psi}{\partial a_\rho} = 0 \qquad (3.2.12)$$

hold. Together, (3.2.10) and (3.2.12) are $N-1$ equations from which may be calculated the functions $a_\rho(x^i)$. These, on substitution into (3.2.6), provide a solution of the partial differential equation. We shall leave to the reader the examination of equations (3.2.10)–(3.2.12) when the number of relations among the a_ρ is greater than one. The result is the same in each case, namely, that the elimination of the a_ρ from a number of relations of the form (3.2.10) and (3.2.12), together with (3.2.6), leads to the solution function $u(x^i)$.

Given a complete integral (3.2.3), we may express the arbitrary constants therein as independent functions of an equal number of new parameters, thus obtaining a complete integral of a different form. It is clear that there are infinitely many ways of expressing a complete integral—these correspond to different ways of parametrizing those surface elements at each point which satisfy the differential equation.

Since it is geometrically evident that the envelope of a family of integral surfaces is also an integral surface, it follows that the surfaces obtained by eliminating some or all of the a_ρ from (3.2.6) with the help of equations of the form

$$\frac{\partial \varphi}{\partial a_\rho} = 0$$

are integral surfaces. Indeed, if we eliminate all of the a_ρ in this way, we obtain a single (possibly non-existent) integral known as the singular integral. Again it follows from the theory of the characteristic strips that the envelope of a family of integral surfaces meets each surface throughout the length of a common characteristic strip. In the following section we shall pursue some consequences of this fact.

Exercise 1. For an equation $F(x^i, u, p_i) = 0$, in which u appears explicitly, what modifications are needed in the definition of a complete integral?

Exercise 2. Show that an N-parameter family of surfaces $\varphi(x^i, a_\rho) + c = 0$, which satisfy the condition (3.2.4), determine a partial differential equation (3.2.2) uniquely.

Exercise 3. Find complete integrals of the following equations in two variables:

(a) $F(x^1, p_1, p_2) = 0$,
(b) $a(x^1)p_1^2 + b(x^2)p_2^2 = c$ ($c =$ constant).

3.3. The Hamilton–Jacobi theorem. We have seen that the general equation of the first order can be reduced to the form

$$F(x^i, p_i) = 0,   \tag{3.3.1}$$

possibly at the cost of increasing by one the number of independent variables. Now the equations of the characteristic strips corresponding to (3.3.1) simplify from the general case (3.1.10). Indeed, they fall naturally into two groups, of $2N$ and of 1 in number. We have, first,

$$\frac{dx^i}{dt} = \frac{\partial F}{\partial p_i}; \qquad \frac{dp_i}{dt} + \frac{\partial F}{\partial x^i} = 0.   \tag{3.3.2}$$

Since F contains only the x^i and p_i, this set of $2N$ equations is 'self-contained' and may be integrated separately from the one remaining equation

$$\frac{du}{dt} = p_i \frac{\partial F}{\partial p_i}.   \tag{3.3.3}$$

When solutions of (3.3.2) are supplied in (3.3.3), the dependent variable u may be determined along the characteristic strip.

The system (3.3.2) has the form of Hamilton's equations for a dynamical system whose 'Hamiltonian function' is $F(x^i, p_i)$. It is therefore of great interest to reverse the viewpoint which we have previously held, and to inquire whether a knowledge of solutions of the partial differential equation will enable us to determine solutions of the system (3.3.2) of ordinary differential equations. We shall discover that this can indeed be done, provided that we know a complete integral of the partial differential equation. The possibility of deriving solutions (motions of a dynamical system) from the complete integral of a related partial differential equation was first realized by Jacobi. The $2N$ variables x^i and p_i appear on an equal footing as dependent variables in the Hamiltonian system (3.3.2). In the space of $2N$ dimensions of these variables—the phase space of Hamiltonian dynamics—a solution of (3.3.2) gives rise to a curve, an integral curve which is the path of a motion of the dynamical system. In this phase space, the system (3.3.2) is a system of ordinary differential equations of the first order; thus a uniquely determined solution passes through each point. Along any such curve there holds the first integral

$$F(x^i, p_i) = \text{constant.} \qquad (3.3.4)$$

This is the energy integral which asserts that the total energy of the dynamical system is conserved, or that the path lies on a surface of constant energy in the phase space. Since we are concerned with the equation (3.3.1), we shall be restricted to the surface of zero energy—a surface of $2N-1$ dimensions on which is traced a $2N-2$ parameter family of integral curves of the system (3.3.2).

Let

$$u = \varphi(x^i, a_\rho) + c \qquad (3.3.5)$$

be a complete integral of (3.3.1), where

$$p_i = \frac{\partial \varphi}{\partial x^i}: \qquad (3.3.6)$$

Then the equations

$$\frac{\partial \varphi}{\partial a_\rho} = b^\rho \qquad (3.3.7)$$

($b^\rho = $ constant; $\rho = 1, ..., N-1$), together with (3.3.6), determine a $(2N-2)$-parameter family of integral curves in the phase space of the Hamiltonian system (3.3.2).

We see that (3.3.6) and (3.3.7) together are $2N-1$ equations among the $2N$ variables x^i, p_i; so that they define curves (one-dimensional loci) which depend on the $2N-2$ parameters a_ρ and b^ρ. Now these curves all lie on the surface $F(x^i, p_i) = 0$ 'of zero energy' in the $2N$-dimensional space, because (3.3.5) is a complete integral of (3.3.1). This means just that the a_ρ can be eliminated from the N equations (3.3.6) and that the single relation among the x^i and p_i remaining after this elimination is precisely $F(x^i, p_i) = 0$. Let τ be a parameter which varies along the curves defined by (3.3.6) and (3.3.7). Thus we may express x^i and p_i as functions of the $2N-1$ quantities a_ρ, b^ρ, and τ:

$$x^i = x^i(a_\rho, b^\rho, \tau), \qquad p_i = p_i(a_\rho, b^\rho, \tau). \qquad (3.3.8)$$

We must show that, for a suitable choice of τ, these functions satisfy (3.3.2).

Differentiate (3.3.7) with respect to τ; since b^ρ is fixed along any one integral curve, we see that

$$\frac{\partial^2 \varphi}{\partial a_\rho \partial x^i} \frac{dx^i}{d\tau} = 0. \qquad (3.3.9)$$

Now differentiate the partial differential equation (3.3.1) with respect to a_ρ, the p_i being regarded as functions of the a_ρ defined by (3.3.6). We find

$$\frac{\partial^2 \varphi}{\partial a_\rho \partial x^i} \frac{\partial F}{\partial p_i} = 0. \qquad (3.3.10)$$

But the rank of the matrix of coefficients in both (3.3.9) and (3.3.10), namely, $(\partial^2 \varphi / \partial a_\rho \partial x^i)$, is $N-1$; there must subsist a proportionality

$$\frac{dx^i}{d\tau} = \rho \frac{\partial F}{\partial p_i} \qquad (3.3.11)$$

$(\rho = \rho(\tau); \ i = 1,...,N)$ between the 'solutions' of these two equations. Let us now choose

$$t = \int \cdot \rho \, d\tau; \qquad (3.3.12)$$

it is evident that (3.3.11) now takes the form of the first set of the equations (3.3.2). The second set may be found as follows. Differentiate (3.3.6) with respect to τ:

$$\frac{dp_i}{d\tau} = \frac{\partial^2 \varphi}{\partial x^i \partial x^k} \frac{dx^k}{d\tau}, \qquad (3.3.13)$$

and differentiate (3.3.1) with respect to x^i: thus

$$\frac{\partial F}{\partial x^i} + \frac{\partial F}{\partial p_k} \frac{\partial p_k}{\partial x^i} = \frac{\partial F}{\partial x^i} + \frac{\partial^2 \varphi}{\partial x^i \partial x^k} \frac{\partial F}{\partial p_k} = 0. \qquad (3.3.14)$$

We find, therefore, that

$$\frac{dp_i}{d\tau} = \frac{\partial^2 \varphi}{\partial x^i \partial x^k} \frac{dx^k}{d\tau} = \frac{\partial^2 \varphi}{\partial x^i \partial x^k} \rho \frac{\partial F}{\partial p_k} = -\rho \frac{\partial F}{\partial x^i}, \qquad (3.3.15)$$

and, in view of (3.3.12), the second set of Hamiltonian equations (3.3.2) is surely satisfied. This completes the proof of the Hamilton–Jacobi theorem.

The following example shows the use of this theorem in determining the orbit of a particle moving in a plane under a central (gravitational) force whose potential $V(r)$ depends only on the distance r from the origin in the (r, θ) plane. The differential equations are of the Hamiltonian type, where

$$F = H = \tfrac{1}{2}\{p_r^2 + (1/r^2)p_\theta^2\} + V(r).$$

represents the 'total energy'. We have thus to find a complete integral of the partial differential equation

$$p_r^2 + (1/r^2)p_\theta^2 + 2V(r) = C \quad (C = \text{constant}).$$

Such an integral can be found by assuming a solution in the separated form

$$u = f(r) + g(\theta),$$

whence $$[f'(r)]^2 + (1/r^2)[g'(\theta)]^2 + 2V(r) = C,$$

so that $g'(\theta) = \beta = \text{constant}$, and

$$f'(r) = [C - 2V(r) - (\beta^2/r^2)]^{\frac{1}{2}}.$$

Thus $$u = \varphi(r, \theta) = \beta\theta + \int^r [C - 2V(\rho) - (\beta^2/\rho^2)]^{\frac{1}{2}} \, d\rho + \gamma,$$

and this is a complete integral. From the theorem we see that the characteristic curves are given by

$$\frac{\partial \varphi}{\partial \beta} = \theta_0,$$

that is, by $$\theta - \theta_0 = \beta \int^r \frac{d\rho}{\rho^2 [C - 2V(\rho) - (\beta^2/\rho^2)]^{\frac{1}{2}}}.$$

This is the equation of the orbit. Under Newtonian gravitation, $V(r) = -1/r$ and the integral can be evaluated as an inverse trigonometric function.

In dynamical applications of this result it is usually required to find the motion of a system, rather than just the path. Let us note that if the time is taken on an equal footing with the other coordinates, and the configuration and phase space correspondingly enlarged, then the form of the result given above will apply, since the motion in configuration space is given by the path alone in the enlarged configuration space.

Exercise 1. Verify (3.3.4).

Exercise 2. If (3.3.2) defines a motion of a fluid in the phase space, show that the (Cartesian) divergence of the flow is zero (Liouville's theorem).

Exercise 3. Find the motion of a particle in a central field of force as in the example above, given that the function F is now

$$p_t + \tfrac{1}{2}\{p_r^2 + (1/r^2)p_\theta^2\} + V(r).$$

Exercise 4. Find a complete integral of

$$F = \tfrac{1}{2}p_x^2 + \tfrac{1}{2}p_y^2 + gy - h = 0,$$

where h and g are constants. Interpret the physical meaning of the equation and apply the Hamilton–Jacobi theorem to find the paths.

3.4. Involutory systems. The method of constructing solutions by means of characteristic strips as outlined in § 1 of this chapter is, in principle, all that is needed to integrate a single partial differential equation of the first order. However, we shall now consider systems of such equations (having always one dependent variable u), and it is necessary to develop a rather different technique, which will also result in a direct method for finding a complete integral of a given single equation by purely formal means. A good deal of complication can be avoided if we restrict the treatment to equations which do not contain the dependent variable explicitly. No loss of generality is entailed by this restriction, as we have seen.

Consider the system of r equations, where $r < N$:

$$F_\rho(x^i, p_i) = 0 \quad (\rho = 1, ..., r). \tag{3.4.1}$$

The problem of finding solutions can be expressed in the following way: *find N functions $p_i = f_i(x^k)$, such that* (3.4.1) *hold identically in the variables x^k; and such that the total differential equation*

$$du = f_i(x^k)\, dx^i \tag{3.4.2}$$

is completely integrable. The conditions of integrability of (3.4.2) are

$$\frac{\partial f_i}{\partial x^k} = \frac{\partial f_k}{\partial x^i}, \tag{3.4.3}$$

since the functions f are independent of u. Indeed, any integral $u = u(x^k)$ of the total differential equation certainly satisfies

$$\frac{\partial u}{\partial x^i} \equiv p_i = f_i(x^k),$$

and therefore is a solution of the system (3.4.1). From a purely theoretical point of view, the integration of (3.4.2), which involves only a quadrature, is less difficult than the solution of (3.4.1) by other means.

Bearing in mind the corresponding facts regarding linear equations, we may ask what additional first-order equations independent of (3.4.1) are necessarily satisfied by all solutions of this system. Let $u(x^k)$ be a common solution, then

$$\frac{\partial F_\rho}{\partial x^i} + \frac{\partial F_\rho}{\partial p_k} \frac{\partial p_k}{\partial x^i} = 0. \tag{3.4.4}$$

If we multiply by $\partial F_\rho/\partial p_i$ and contract over the index i, we have

$$\frac{\partial F_\rho}{\partial x^i} \frac{\partial F_\sigma}{\partial p_i} = -\frac{\partial F_\rho}{\partial p_k} \frac{\partial F_\sigma}{\partial p_i} \frac{\partial p_k}{\partial x^i} = -\frac{\partial F_\rho}{\partial p_k} \frac{\partial F_\sigma}{\partial p_i} \frac{\partial p_i}{\partial x^k} = \frac{\partial F_\sigma}{\partial x^i} \frac{\partial F_\rho}{\partial p_i}, \tag{3.4.5}$$

again using (3.4.4) with σ replacing ρ. It therefore follows that the 'Poisson bracket'

$$(F_\sigma, F_\rho) = \frac{\partial F_\rho}{\partial x^i} \frac{\partial F_\sigma}{\partial p_i} - \frac{\partial F_\sigma}{\partial x^i} \frac{\partial F_\rho}{\partial p_i} \tag{3.4.6}$$

must vanish as a consequence of the vanishing of F_ρ and F_σ.

Exercise. Show that (F, G) is an invariant under transformations of the independent variables x^i.

The common integrals of the system $F_\rho = 0$ must therefore satisfy the further equations

$$(F_\rho, F_\sigma) = 0, \tag{3.4.7}$$

some of which may be independent of the original set of equations. Let us adjoin to the given system all of the new equations so formed. Continuing to form Poisson brackets for the enlarged system, and to adjoin those new equations which are independent, we must eventually reach one of the following alternatives. Either there arise more than N independent equations, in which case the system has no solutions, or else there results a system of $p < N$ equations such that all Poisson brackets of the form (3.4.6) vanish, either identically or as a consequence of the given set of p equations.

These enlarged systems, which we shall now consider, correspond to complete systems of linear equations. As in the linear case, it is always possible to transform the equations in such a way that the Poisson brackets vanish identically. Such systems, which correspond to the linear Jacobian systems, are known as involutory systems. Here is a proof that an enlarged system of $p < N$ equations can be transformed to an involutory system. Select p of the derivatives p_i, say p_α ($\alpha = 1,...,p$), such that the equations $F_\rho = 0$ can be solved for the p_α in terms of the x^k and p_ρ ($\rho = p+1,...,N$),

$$p_\alpha = f_\alpha(x^k, p_\rho). \tag{3.4.8}$$

These p equations are equivalent, in the functional sense, to the given enlarged system. Now the Poisson brackets

$$(p_\alpha - f_\alpha, p_\beta - f_\beta) \qquad (3.4.9)$$

for $\alpha, \beta = 1,...,p$, when expanded, contain none of the p_α. Thus the equations obtained by equating these brackets to zero cannot be consequences of (3.4.8). Yet the brackets (3.4.9) vanish, since if they did not, there would subsist a further relation among the p_ρ ($\rho = p+1,...,N$), contrary to the hypothesis that the system $F_\rho = 0$ ($\rho = 1,...,p$) contained itself all of the conditions on the p_i which can be found by differentiation and elimination. Therefore, the brackets (3.4.9) must be identically zero in all of the variables x^i and p_ρ ($\rho = p+1,...,N$). That is, the system (3.4.8) is involutory.

Now the alternative formulation which we gave for the integration of a system of first-order equations can be connected with the properties of involutory systems which contain exactly N equations. In fact, we can show that such a system can be integrated by a quadrature. Let

$$F_\rho(x^i, p_i) = 0 \qquad (3.4.10)$$

be an involutory system with $\rho = 1,...,N$. Then

$$(F_\sigma, F_\rho) = 0 \qquad (3.4.11)$$

for $\rho, \sigma = 1,...,N$, and, since the F_ρ are independent functions of the p_i, we have

$$\frac{\partial(F_\rho)}{\partial(p_i)} \equiv \frac{\partial(F_1,...,F_N)}{\partial(p_1,...,p_N)} \neq 0. \qquad (3.4.12)$$

If now the equations $\qquad F_\rho(x^i, p_i) = a_\rho \qquad (3.4.13)$

are solved for the p_i as functions of the x^i and a_ρ ($i, \rho = 1,...,N$):

$$p_i = f_i(x^k, a_\rho) \qquad (3.4.14)$$

then the total differential equation

$$du = p_i \, dx^i = f_i(x^k, a_\rho) \, dx^i \qquad (3.4.15)$$

is completely integrable.

This statement is not hard to prove. Differentiate the condition (3.4.13) with respect to x^i, and obtain

$$\frac{\partial F_\rho}{\partial x^i} + \frac{\partial F_\rho}{\partial p_k} \frac{\partial p_k}{\partial x^i} = 0.$$

Now multiply by $\partial F_\sigma / \partial p_i$ and subtract from the result a similar equation in which ρ and σ have been interchanged. In view of the definition of the

Poisson bracket, the resulting condition is seen to be

$$(F_\sigma, F_\rho) + \frac{\partial F_\sigma}{\partial p_i}\frac{\partial F_\rho}{\partial p_k}\left(\frac{\partial p_k}{\partial x^i} - \frac{\partial p_i}{\partial x^k}\right) = 0. \qquad (3.4.16)$$

But, by hypothesis, the Poisson bracket is zero. We should like to prove now that the matrix **B** of quantities

$$\frac{\partial p_k}{\partial x^i} - \frac{\partial p_i}{\partial x^k} \equiv \frac{\partial f_k}{\partial x^i} - \frac{\partial f_i}{\partial x^k} \qquad (i, k = 1,...,N),$$

is identically zero. To accomplish this, note that (3.4.16) takes the matrix form

$$\mathbf{ABA} = \mathbf{0},$$

where **A** is the matrix of the quantities $\partial F_\rho/\partial p_i$. Now, in view of (3.4.12), **A** is non-singular, and has therefore an inverse $\mathbf{A}^{-1}$. If we multiply our matrix equation on right and left by $\mathbf{A}^{-1}$, we find just the result $\mathbf{B} = \mathbf{0}$; so that, finally,

$$\frac{\partial f_k}{\partial x^i} = \frac{\partial f_i}{\partial x^k}. \qquad (3.4.17)$$

These are the conditions of integrability for (3.4.15).

Let us suppose that (3.4.15) can be integrated explicitly; thus let

$$u = \varphi(x^i, a_\rho) = \int f^i(x^k, a_\rho)\, dx^i + c. \qquad (3.4.18)$$

Since

$$\frac{\partial \varphi}{\partial x^i} = f_i(x^k, a_\rho), \qquad (3.4.19)$$

we see that the determinant

$$\left|\frac{\partial^2 \varphi}{\partial x^i \partial a_\rho}\right| = \left|\frac{\partial f_i}{\partial a_\rho}\right| = \left|\frac{\partial p_i}{\partial F_\rho}\right| \qquad (3.4.20)$$

is just the inverse determinant $|\mathbf{A}|^{-1}$, and is therefore different from zero. Thus (3.4.18) is actually a complete integral.

Exercise 1. Work out explicitly the conditions $(F_\sigma, F_\rho) = 0$, when F_ρ is linear in the p_i:

$$F_\rho = b^i_\rho p_i, \qquad b^i_\rho = b^i_\rho(x^k).$$

Exercise 2. Let t be the parameter along the characteristic strips of the partial differential equation $F(x^i, p_i) = 0$. If $G(x^i, p_i)$ is any function of the same variables, show that along the characteristic curves we have

$$\frac{dG}{dt} = (F, G).$$

Exercise 3. If $F_\rho(x^i, u, p_i) = 0$ $(\rho = 1,...,p)$ where the F_ρ contain u explicitly, show that

$$\frac{\partial F_\rho}{\partial p_i}\frac{dF_\sigma}{dx^i} - \frac{\partial F_\sigma}{\partial p_i}\frac{dF_\rho}{dx^i} = 0 \quad (\rho, \sigma = 1,...,p),$$

where the differentiation with the straight d is defined as in § 1.3.

Exercise 4. Show that the system in three variables

$$x^1 p_2 p_3 = x^2 p_3 p_1 = x^3 p_1 p_2 = x^1 x^2 x^3$$

gives rise to an involutory system, and integrate the corresponding equation (3.4.15).

3.5. Jacobi's integration method. Our study of involutory systems has shown, first, that any system of $r \leqslant N$ equations which are compatible can be extended to an involutory system of p equations, with $r \leqslant p \leqslant N$, and, second, that a system of N involutory equations leads to a completely integrable total differential equation. The integration method of Jacobi consists of finding $N-p$ additional relations which, when added to the equations of an involutory system of p equations, yield an involutory system of N equations which contains $N-p$ arbitrary constants.

An essential feature of this method is the use of the following identity in the Poisson brackets of any three suitable functions f, g, h, of the variables x^i and p_i:

$$((f,g),h) + ((g,h),f) + ((h,f),g) = 0. \tag{3.5.1}$$

This identity, which bears the name of Poisson, can be proved by expanding the left-hand side in full and observing that all terms cancel. An indirect proof, which is a little shorter, depends on the observation that each term on the left, when it is expanded, contains a second derivative of one of the functions f, g, or h, multiplied by first derivatives of the other two functions. Let us show that all terms with second derivatives of, say, f, disappear. These terms are contained only in

$$((f,g),h) + ((h,f),g) = (h,(g,f)) - (g,(h,f)) = Y(X(f)) - X(Y(f)),$$
$$\tag{3.5.2}$$

where we have set

$$(g,f) = X(f), \qquad (h,f) = Y(f).$$

Thus X and Y are linear differential operators of the kind studied in § 2.3, and the expression (3.5.2) is the commutator expression of X and Y operating on f. But we proved that this expression contains no second derivatives of f. Thus the left-hand side of (3.5.1) contains no second derivatives of f, or, by symmetry, of g or of h, and so it must be zero. This proves the Poisson identity.

Turning to the Jacobi method proper, let us consider the involutory system

$$F_\rho(x^i, p_i) = 0 \quad (\rho = 1,...,p < N). \tag{3.5.3}$$

Thus

$$(F_\sigma, F_\rho) = 0 \quad (\rho, \sigma = 1,...,p). \tag{3.5.4}$$

Now any function U of the x^i and p_i which is in involution with the given functions F_ρ satisfies the equations

$$X_\rho(U) = (F_\rho, U) = 0. \tag{3.5.5}$$

This is a system of p linear homogeneous first-order equations of the type treated in § 2.3, and we shall apply the results established there. First we shall prove that the linear system (3.5.5) is Jacobian. For, in view of (3.5.1), (3.5.4), and (3.5.5), we have identically

$$X_\rho(X_\sigma(U)) - X_\sigma(X_\rho(U)) = (F_\rho, (F_\sigma, U)) - (F_\sigma, (F_\rho, U))$$
$$= (F_\rho, (F_\sigma, U)) + (F_\rho, (U, F_\sigma)) \tag{3.5.6}$$
$$= -(U, (F_\rho, F_\sigma)) = 0.$$

Now, since we have supposed $p < N$, there exists an integral of the system (3.5.5), which we may denote by $F_{p+1}(x^i, p_i)$. Let us now adjoin to the given system (3.5.3) of non-linear equations the additional relation

$$F_{p+1}(x^i, p_i) = a_{p+1} \quad (a_{p+1} = \text{constant}), \tag{3.5.7}$$

thus obtaining an involutory system of $(p+1)$ equations. Each solution of the augmented system is certainly a solution of the original system (3.5.3).

The continuation of this method is obvious. Repeating the process $(N-p)$ times, we find successively $(N-p)$ new functions $F_{p+1}, F_{p+2},..., F_N$ which are in involution with each other and with the functions $F_\rho (\rho = 1,...,p)$. Then finally, according to the result of the previous section, we solve the N relations consisting of (3.5.3) together with

$$F_\rho(x^i, p_i) = a_\rho \quad (\rho = p+1,...,N), \tag{3.5.8}$$

for the p_i as functions of the x^i and the a_ρ. We then substitute the resulting functions

$$p_i = f_i(x^k, a_\rho)$$

in the relation

$$du = p_i \, dx^i,$$

which, as we saw, will necessarily be completely integrable. On integration we find a function

$$u = \varphi(x^i, a_\rho) + c, \tag{3.5.9}$$

which depends on the $(N-p)$ constants a_ρ, and the additional constant c. From (3.4.20) it follows easily that the family of integral surfaces given by

(3.5.9) contains $N-p+1$ independent parameters. Thus (3.5.9) is a complete integral of the given system of p non-linear equations.

Though in principle Jacobi's method is adequate to solve any system of equations of the first order in one dependent variable, it is evidently rather laborious, particularly when p is small compared with N.

For a single equation we can by this method obtain a complete integral containing N independent constants. Let us examine this case a little more closely. Those functions U which are in involution with a given function $F(x^i, p_i)$, and only those, satisfy the linear equation in $2N$ variables

$$\frac{\partial F}{\partial p_i}\frac{\partial U}{\partial x^i} - \frac{\partial F}{\partial x^i}\frac{\partial U}{\partial p_i} = 0. \qquad (3.5.10)$$

Regarded as a linear equation for U, (3.5.10) has the characteristic equations

$$\frac{dx^1}{\dfrac{\partial F}{\partial p_1}} = \frac{dx^2}{\dfrac{\partial F}{\partial p_2}} = \cdots = \frac{dx^N}{\dfrac{\partial F}{\partial p_N}} = \frac{dp_i}{-\dfrac{\partial F}{\partial x^1}} = \cdots = \frac{dp_N}{-\dfrac{\partial F}{\partial x^N}}. \qquad (3.5.11)$$

Suppose that we can find $N-1$ independent first integrals of (3.5.11) which, as functions of x^i and p_i, are in involution with each other, and which, together with the given equation, can be solved for the p_i as functions of the x^i. We can then obtain a complete integral by a quadrature, as before.

For example, any equation in which the independent variables themselves do not appear is of this kind. Let the equation be

$$F(p_i) = 0,$$

then $\partial F/\partial x^i = 0$ and from (3.5.11) it follows that

$$p_i = a_i \qquad\qquad (i = 1,...,N),$$

are independent first integrals. Since the x^i are not present, these integrals are all involutory. Actually, the complete integral so obtained is a linear function,

$$u = \varphi(x^i, a_\rho) = \sum_i a_i x^i + c.$$

This apparently contains one constant in excess; but the numbers a_i must be chosen so that $F(a_i) = 0$.

This concludes our treatment of first-order equations. In Chapter IX it will be seen that the characteristics of non-linear equations of the first order play an important role in the theory of hyperbolic equations of the second order.

Exercise 1. Solve the system in three variables

$$p_1 p_2 p_3 - a^3 x^1 x^2 x^3 = 0,$$

$$p_2 + p_3 - a(x^2 + x^3) = 0.$$

Exercise 2. For the equation $F(u, p_i) = 0$, in which the x^i do not appear, but u does, show that the ratios p_i / p_k are constant along characteristic strips of $F = 0$. Reduce the integration to a problem in ordinary differential equations by showing that u is a function of $\sum a_k x^k$ alone. Integrate $u = p_1 p_2$; $u = p_1 + p_2 + p_3$.

Exercise 3. If $F(x^k, p_k) = \sum_k f_k(x^k, p_k) = 0$, where each term contains only one coordinate x^k and the corresponding derivative p_k, show that $f_k(x^k, p_k) = c_k$ $(k = 1, ..., N)$ provide N independent first integrals along the characteristic strips. Show how a complete integral may be found. Find a complete integral of $p_1 - (x^1)^2 = p_2 + (x^2)^2$.

Exercise 4. Find a complete integral of the Clairaut equation

$$u = \sum_k p_k x^k + f(p_k).$$

If $f(p_k) = [1 + p_1^2 + p_2^2 + ... + p_N^2]^{\frac{1}{2}}$, show that the complete integral represents all hyperplanes in (x^i, u) space at unit distance from the origin.

IV

LINEAR EQUATIONS OF THE SECOND ORDER

THE study of equations of the second order, which we commence in this chapter, will occupy the rest of this book. We shall begin with the geometric foundations of the theory which, for second-order equations, are more elaborate than for equations of the first order. This background is based on Riemannian geometry, which we shall develop as needed, but necessarily from a point of view restricted to our main purpose. A more detailed account of Riemannian geometry may be found in any of the books on tensor calculus listed in the bibliography.

The general linear partial differential equation of the second order in N independent variables x^i, and one dependent variable u, has the form

$$L[u] = a^{ik} \frac{\partial^2 u}{\partial x^i \partial x^k} + \beta^i \frac{\partial u}{\partial x^i} + cu = f,$$

where a^{ik}, β^i, c, and f are given functions of the independent variables x^i, but do not contain u. The summation convention is understood, and applies to the indices i and k. The quantities a^{ik}, β^i, and c all influence the nature of the differential equation, but the major role falls to the coefficients a^{ik} of the second derivatives. In the first section the tensor character of these coefficients is brought out, and the nature of the classification into elliptic, hyperbolic, and parabolic types determined by the a^{ik} is shown. Next we introduce the Riemannian geometry with its concepts of length and volume, for equations of elliptic and hyperbolic type. Then certain basic invariant differential operators are defined, and the operator $L[u]$ is expressed by means of them.

In the third section is derived Green's formula, the principal analytical tool of all our subsequent work. Next, we consider equations with constant coefficients, and construct a 'fundamental' solution for such equations. The attempt to extend this construction to equations having variable coefficients leads us back to Riemannian geometry in the fifth section, in which is briefly set forth the theory of geodesic lines.

The next four chapters centre about the theory of elliptic equations, while the two remaining chapters are devoted to the equations of hyperbolic type.

4.1. Classification; the fundamental tensor. Consider a linear differential operator L, defined by

$$L[u] = a^{ik} \frac{\partial^2 u}{\partial x^i \partial x^k} + \beta^i \frac{\partial u}{\partial x^i} + cu, \qquad (4.1.1)$$

where the indices i and k are understood to be summed from 1 to N, N being the number of independent variables x^i. Here a^{ik}, β^i, and c are the coefficients of this operator, which is defined by the form (4.1.1) in a given system of coordinates x^i. We shall suppose that these coefficients are functions of the N variables x^i. We should like to assign a significance to $L[u]$ which is independent of the particular coordinate system, and we therefore inquire: what law of transformation of the coefficients will permit an invariant significance for the value of $L[u]$? Under transformations of the coordinate variables, we have

$$\frac{\partial u}{\partial x^i} = \frac{\partial u}{\partial x'^m} \frac{\partial x'^m}{\partial x^i}, \qquad (4.1.2)$$

and

$$\frac{\partial^2 u}{\partial x^i \partial x^k} = \frac{\partial^2 u}{\partial x'^m \partial x'^n} \frac{\partial x'^m}{\partial x^i} \frac{\partial x'^n}{\partial x^k} + \frac{\partial^2 x'^m}{\partial x^i \partial x^k} \frac{\partial u}{\partial x'^m}. \qquad (4.1.3)$$

Thus

$$L[u] = a^{ik} \frac{\partial x'^m}{\partial x^i} \frac{\partial x'^n}{\partial x^k} \frac{\partial^2 u}{\partial x'^m \partial x'^n} + \left(a^{ik} \frac{\partial^2 x'^m}{\partial x^i \partial x^k} + \beta^i \frac{\partial x'^m}{\partial x^i} \right) \frac{\partial u}{\partial x'^m} + cu$$

$$= a'^{mn} \frac{\partial^2 u}{\partial x'^m \partial x'^n} + \beta'^m \frac{\partial u}{\partial x'^m} + cu, \qquad (4.1.4)$$

where, in particular, $a'^{mn} = a^{ik} \dfrac{\partial x'^m}{\partial x^i} \dfrac{\partial x'^n}{\partial x^k}.$ (4.1.5)

The corresponding formula for the coefficients β'^m is evident and need not be written down. Also, we observe that the coefficient of u is unchanged. The form of equation (4.1.5) shows us that the coefficients a^{ik} of the second derivatives in $L[u]$ obey the transformation law of a contravariant tensor of the second order. This result we have anticipated by the notation. Indeed, we may say that the operator $L[u]$ defines a contravariant tensor a^{ik}, whose components in any coordinate system are just the coefficients of the second derivatives of u, as they appear in $L[u]$ expressed in that system. For the present let us consider these coefficients a^{ik} apart from the b^i and c, since, as we shall see, the most important properties of $L[u]$ are determined by the tensor a^{ik}. One rather natural simplification which can now be made without loss of generality is that a^{ik} can be taken as symmetric: $a^{ik} = a^{ki}$. Indeed, since the mixed second derivatives of u

are symmetric, the coefficients a^{ik} appear in $L[u]$ in the combination $a^{ik}+a^{ki}$. Our assumption of symmetry means only that the total coefficient is to be shared equally between the two terms in which it appears. This symmetry is preserved under transformations of the coordinates. Thus the operator L determines a symmetric contravariant tensor a^{ik}, which we shall call the fundamental tensor.

Let ξ_i be a covariant vector representing a surface element at a point P. Then the quadratic form

$$Q(\xi) = a^{ik}\xi_i\xi_k, \tag{4.1.6}$$

which is evidently an invariant, is known as the characteristic form relative to our operator $L[u]$. The significance of Q is as follows: let $\varphi(x^i) = 0$ be a surface passing through P, then the coefficient in $L[u]$ of the second derivative of u with respect to φ, that is, across this surface, is equal to $Q(\varphi)$. To see this, let us choose a new system of coordinates of which one is $x'^1 = \varphi(x^i)$. Then the coefficient of $\partial^2u/\partial x'^1\partial x'^1$ in $L[u]$ is

$$a'^{11} = a^{ik}\frac{\partial\varphi}{\partial x^i}\frac{\partial\varphi}{\partial x^k} \equiv Q(\varphi). \tag{4.1.7}$$

Should this coefficient be zero, the surface element $\xi_i \equiv \partial\varphi/\partial x^i$ is said to be characteristic. Evidently the coefficient of the second derivative of u across or transverse to a characteristic surface vanishes, so that $L[u]$ can be calculated at P if the values of u and of the first derivatives of u on the surface $\varphi(x^i) = 0$ are known. In a sense, therefore, $L[u]$ degenerates to an operator of the first order on a characteristic surface.

When $L[u]$ has no characteristic surface elements at P, it is said to be *elliptic*. Clearly $L[u]$ is elliptic at P if and only if the quadratic form $Q(\varphi)$ is definite, and we can without loss of generality assume that in fact $Q(\varphi)$ is positive definite. An elliptic operator $L[u]$ is therefore of the full order two on every surface element at P. It is a standard result of matrix theory that if a quadratic form is definite, the matrix of its coefficients is non-singular and thus possesses an inverse matrix.

The matrix of coefficients a^{ik} may, however, be non-singular without the characteristic form being definite. In this case of a non-singular, indefinite characteristic form the operator $L[u]$ is said to be *hyperbolic*. Finally, if the matrix of coefficients is singular, so that the determinant $|a^{ik}|$ is zero, $L[u]$ is said to be *parabolic*. An operator is parabolic if the differential operations of the second order in it can be expressed in terms of fewer than N variables. We shall first demonstrate that this classification into the three types is independent of the system of coordinates in which the coefficients a^{ik} are expressed. The determinant of the coefficients a^{ik} has, in view of the law of determinant multiplication, the following transformation law:

$$|a'^{mn}| = \left|\frac{\partial x'^m}{\partial x^i}\frac{\partial x'^n}{\partial x^k}a^{ik}\right| = \left|\frac{\partial x'^m}{\partial x^i}\right|\left|\frac{\partial x'^n}{\partial x^k}a^{ik}\right|$$

$$= \left|\frac{\partial x'^m}{\partial x^i}\right|\left|\frac{\partial x'^n}{\partial x^k}\right||a^{ik}| = \left|\frac{\partial x'}{\partial x}\right|^2|a^{ik}|. \qquad (4.1.8)$$

Under a non-singular transformation, the functional determinant $|\partial x'/\partial x|$ is not zero; hence if $|a^{ik}|$ is different from zero in one coordinate system, so is it in all others. Conversely, the vanishing of $|a^{ik}|$ in one system must imply its vanishing in every other system. That is, an equation which is parabolic in one coordinate system is parabolic in every system. Again, regarding the distinction between elliptic and hyperbolic equations, we see that the definite or indefinite character of the invariant quadratic form $Q(\varphi)$ is certainly independent of particular coordinate systems. This establishes the invariant nature of the classification.

Since the coefficients a^{ik} may depend on the point P, an operator $L[u]$ may be of different types at different points of the space of the coordinate variables. Let us note, however, that if there exists some coordinate system, in which the a^{ik} are all constants, then the type of $L[u]$ must be the same at every point.

The significance of the classification into elliptic, hyperbolic, and parabolic types is profound, and will fully emerge only as we proceed. We shall now show how the characteristic form may be expressed in a simplified standard way in the various cases. It is well known that by a linear transformation any quadratic form can be expressed as a sum of terms

$$Q(\eta) = \sum_{i=1}^{N} \epsilon_i \eta_i^2, \qquad (4.1.9)$$

where the ϵ_i are all either $+1$, 0, or -1. Now the quadratic form (4.1.9) can be positive definite, corresponding to an elliptic operator, only if all of the ϵ_i are $+1$. Also, the form $Q(\eta)$ is non-singular only if none of the ϵ_i is zero, for then the determinant of (4.1.9) would vanish. Thus a hyperbolic operator corresponds to the case where some of the ϵ_i are $+1$, some -1. If one only of the ϵ_i is positive (or one only negative), then $L[u]$ is called normal hyperbolic; otherwise $L[u]$ is ultrahyperbolic. Of the various types of hyperbolic equations, those which are normal hyperbolic are most frequent in applications and are the best understood.

The classification can also be established by means of the latent or characteristic roots of the matrix A of the coefficients a^{ik}, that is, the roots of the polynomial equation in λ:

$$|A - \lambda I| = 0.$$

Since A is symmetric, the roots $\lambda_1, \ldots, \lambda_N$ are all real. The product of these roots is in fact equal to the determinant $|A| = |a^{ik}|$, so that $L[u]$ is parabolic if and only if one or more of the λ_i are zero. If all of the λ_i are positive, $L[u]$ is elliptic; if one of the roots differs in sign from the rest, and none are zero, then $L[u]$ is of normal hyperbolic type.

Exercise 1. In what regions are the following operators elliptic, hyperbolic, or parabolic:

$$\frac{\partial^2 u}{(\partial x^1)^2} + x^1 \frac{\partial^2 u}{(\partial x^2)^2} + \frac{\partial u}{\partial x^2}; \qquad \frac{\partial^2 u}{\partial x^1 \partial x^2} + x^3 \frac{\partial^2 u}{\partial x^3 \partial x^4}.$$

Exercise 2. Find the characteristic *curves* of the equation in two variables

$$\frac{\partial^2 u}{\partial x^2} + y \frac{\partial^2 u}{\partial y^2} = 0$$

in the region $y < 0$. Determine the form of the equation when new independent variables $\xi = x$, $\eta = 2y^{\frac{1}{2}}$ are introduced in the region $y > 0$.

Exercise 3. Find the characteristic curves of

$$\frac{\partial^2 u}{\partial x^2} + x \frac{\partial^2 u}{\partial y^2} = 0$$

in the region $x < 0$. Determine the form of the equation in new variables $\xi = \frac{3}{2}y$, $\eta = -x^{\frac{3}{2}}$ defined for $x > 0$.

Exercise 4. If u and v satisfy the system

$$\frac{\partial v}{\partial x} = a(x, y)\frac{\partial u}{\partial x} + b(x, y)\frac{\partial u}{\partial y}, \qquad \frac{\partial v}{\partial y} = c(x, y)\frac{\partial u}{\partial x} + d(x, y)\frac{\partial u}{\partial y},$$

show that u and v each satisfy linear second-order equations of type determined by the sign of $(a - d)^2 + 4bc$.

4.2. Riemannian geometry.

The equations which we plan to study in the following chapters are elliptic and normal hyperbolic. These two types have the greatest interest in application, and in consequence have been investigated most thoroughly. We shall now attempt to provide a suitable geometric background for the study of these equations. The differential operator $L[u]$ defines, and so places at our disposal, a symmetric contravariant tensor a^{ik}, which in these two cases is non-singular:

$|a^{ik}| \neq 0$. Such a tensor is just the raw material needed to define a Rie-mannian metric on the hitherto amorphous space of the independent variables.

Since a^{ik} is non-singular, the matrix (a^{ik}) possesses an inverse matrix, also symmetric, which we may denote by (a_{ik}). This notation anticipates what we shall now prove, that the a_{ik} are components of a covariant tensor. Indeed, the definition of the a_{ik} is provided by the N^2 component equations

$$a^{ik}a_{kj} = \delta^i_j,\tag{4.2.1}$$

where δ^i_j is the Kronecker delta, equal to unity when $i = j$ and zero other-wise. The Kronecker delta δ^i_j is also a mixed tensor of the second order. In each coordinate system the a_{ik} are uniquely determined by (4.2.1). Now let $\bar{a}_{ik}$ be that covariant tensor which, in one given coordinate system, coincides with a_{ik}. Then, the tensor equation

$$a^{ik}\bar{a}_{kj} = \delta^i_j$$

holds in the given system, and so must also hold in all other systems. Thus the $\bar{a}_{ik}$ satisfy (4.2.1) in all coordinate systems and so, in each system, $\bar{a}_{ik} = a_{ik}$. Thus the a_{ik} enjoy the transformation property of a covariant tensor. The two tensors a^{ik}, a_{ik} are said to be conjugate.

With these conjugate tensors we can set up a correspondence between covariant and contravariant vectors, as follows. If b_i is a covariant vector, then b^i, defined by

$$b^i = a^{ik}b_k\tag{4.2.2}$$

is a contravariant vector. Let us refer to b^i as the vector associate to b_i. In view of (4.2.1) we have

$$b_i = a_{ik}b^k,\tag{4.2.3}$$

so that b_i may be regarded as associate to b^i. This process of 'raising' or 'lowering' indices may be extended to tensors of any order. In particular, we may now speak of a covariant vector b_i as the covariant representation of a vector **b**. The contravariant components of **b** are then defined as above. A vector is given, then, by either of its two representations, the covariant components, or the contravariant.

Exercise. Show that a_{ik} and a^{ik} are associate tensors, and find mixed components for a_{ik}.

Let $\varphi(x^j) = c$ denote a surface; or, as c varies, a family of surfaces. The partial derivatives $\partial\varphi/\partial x^j$ which transform covariantly may now be regarded as covariant components of the gradient vector, written $\nabla\varphi$. The contravariant components of the gradient are $a^{ik}(\partial\varphi/\partial x^k)$; if dt is an infinitesimal, then

$$dx^i = a^{ik}\frac{\partial\varphi}{\partial x^k}\,dt$$

is an infinitesimal displacement whose direction will be known as the *normal* to the surface $\varphi = c$. Thus, a covariant vector, representing a surface element, has an associate contravariant vector which defines a direction. This direction we define as the *normal* to the surface element.

The length, or magnitude, of a vector **b** may now be *defined* as the square root of the invariant

$$\mathbf{b}^2 = a_{ik}b^ib^k = a^{ik}b_ib_k = b^ib_i. \tag{4.2.4}$$

This definition of length reduces to the Cartesian definition in Euclidean space where the a^{ik} are equal to zero or unity according as i is different from or equal to k. As a consequence of this definition, an infinitesimal vector ds with contravariant components dx^i has length

$$ds^2 = a_{ik}\,dx^idx^k. \tag{4.2.5}$$

This equation defines a metric, that is, a scheme for assigning a numerical measure to arc lengths, volumes, and other geometrical objects. Since ds^2 is an invariant, this measure is independent of particular coordinate systems as all geometric quantities must be. Since (4.2.5) defines the length of a small arc or line segment, it is called the line element. The coefficients a_{ik} in this fundamental quadratic form (4.2.5) are functions of the coordinates in each coordinate system. This concept of a line element and a geometry more general than the Euclidean based on it, was suggested by Riemann, whose name it bears.

Note that the line element is expressed by means of the covariant components a_{ik} of the fundamental tensor a^{ik} given by the differential equation. From our present point of view we might regard the contravariant components, given first, as the basic ones. In many physical problems, however, the conventions of physics define a metric, and it then is found that partial differential equations whose solutions are desired have as coefficients the components of the associate contravariant tensor a^{ik}.

Suppose that $L[u]$ is elliptic. Then the quadratic form

$$Q(\xi) = a^{ik}\xi_i\xi_k$$

is definite; and, we may suppose, positive definite. Then the length of every vector is positive, and so, therefore, the line element is positive definite. *An elliptic differential equation defines a positive definite Riemannian metric.*

If, however, $L[u]$ is hyperbolic, $Q(\xi)$ assumes both positive and negative values, and there are characteristic surface elements for which $Q(\xi)$ vanishes. If $\xi^i = a^{ik}\xi_k$, then

$$Q(\xi) = a_{ik}\xi^i\xi^k,$$

and it follows that $ds^2 = 0$ along a line which is normal to a characteristic surface element. Such a displacement, with $ds = 0$, is called a null displacement, its direction a null direction.

A measure of volume can be defined from the Riemannian metric. This measure reduces to the usual one in Euclidean space, and also gives the usual definition of area for surfaces when these are regarded as Riemannian spaces whose metric is defined by the metric in the Euclidean space. To derive such a measure of volume, we must associate with each 'solid element' $dx^1 \dots dx^N$ an invariant quantity of the same order of smallness. The formula used in the transformation of multiple integrals, namely,

$$dx'^1 \dots dx'^N = \left| \frac{\partial x'}{\partial x} \right| dx^1 \dots dx^N,$$

shows how the element $dx^1 \dots dx^N$ itself transforms. Here the determinant is the Jacobian of the coordinate transformation. Now the covariant tensor a_{ik} satisfies the law of transformation

$$a'_{mn} = \frac{\partial x^i}{\partial x'^m} \frac{\partial x^k}{\partial x'^n} a_{ik},$$

and from this we see as in (4.1.8) that

$$|a'_{mn}| = \left| \frac{\partial x}{\partial x'} \right|^2 |a_{ik}|.$$

Thus, if we denote $|a_{ik}|$ by a, we have

$$\sqrt{a'}\, dx'^1 \dots dx'^N = \sqrt{a} \left| \frac{\partial x}{\partial x'} \right| \left| \frac{\partial x'}{\partial x} \right| dx^1 \dots dx^N = \sqrt{a}\, dx^1 \dots dx^N,$$

since the determinants are reciprocals. Thus

$$dV = \sqrt{a}\, dx^1 \dots dx^N \qquad (4.2.6)$$

is an invariant of the kind we seek, and so may be called the volume element.

A Riemannian metric enables us to define certain special differential operators in an invariant way. These quantities have a geometrical significance similar to the differential operators used in Cartesian vector calculus, and are evidently generalizations of these operators. We have already mentioned the gradient vector, and the operator ∇ associated to it.

Thus the vector $\nabla\varphi$ has covariant components $\partial\varphi/\partial x^i$ and contravariant components $a^{ik}(\partial\varphi/\partial x^k)$. Its squared length we shall write

$$(\nabla\varphi)^2 = a^{ik}\frac{\partial\varphi}{\partial x^i}\frac{\partial\varphi}{\partial x^k} = Q(\varphi). \qquad (4.2.7)$$

This quadratic form, known also as the first differential parameter of Beltrami, is sometimes denoted by $\Delta_1(\varphi,\varphi)$. If φ and ψ are two scalars, we have for instance

$$\nabla\varphi\cdot\nabla\psi = a^{ik}\frac{\partial\varphi}{\partial x^i}\frac{\partial\psi}{\partial x^k} = \Delta_1(\varphi,\psi)$$

as scalar product. We see also that $(\nabla\varphi)^2$ is zero only if $\varphi = c$ is a characteristic surface. For any surface which is not characteristic, we can define a unit normal $\mathbf{n}$, a vector having the direction of $\nabla\varphi$ and unit magnitude. Thus the N quantities

$$n_k = \frac{1}{\sqrt{\{(\nabla\varphi)^2\}}}\frac{\partial\varphi}{\partial x^k} \qquad (4.2.8)$$

are covariant components of the normal. With the aid of the unit normal we can also define an operation of directional differentiation along the normal. If f is any function, we set

$$\frac{\partial f}{\partial n} = \nabla f\cdot\mathbf{n} = \frac{\nabla f\cdot\nabla\varphi}{\sqrt{\{(\nabla\varphi)^2\}}} = \frac{a^{ik}\dfrac{\partial\varphi}{\partial x^i}\dfrac{\partial f}{\partial x^k}}{\sqrt{\{(\nabla\varphi)^2\}}}; \qquad (4.2.9)$$

this is the *normal derivative*.

Next we shall construct an operator analogous to the 'divergence' of vector calculus. For this purpose we require a formula for the derivative of a Jacobian determinant. To form the derivative of any determinant, we take the derivative of each element in turn, multiply by the corresponding cofactor, and sum over all elements. In the Jacobian $|\partial x'/\partial x|$, the cofactor is the corresponding term in the inverse determinant $|\partial x/\partial x'|$, multiplied by $|\partial x'/\partial x|$ itself. Thus, we have

$$\frac{\partial}{\partial x^k}\left|\frac{\partial x'}{\partial x}\right| = \left|\frac{\partial x'}{\partial x}\right|\frac{\partial x^q}{\partial x'^p}\frac{\partial^2 x'^p}{\partial x^q\partial x^k}, \qquad (4.2.10)$$

summation being taken over the indices p and q.

We shall require the divergence of a vector $\mathbf{b}$ to be an invariant which would reduce, if Cartesian coordinates were possible, to the Cartesian divergence

$$\sum_i\frac{\partial b^i}{\partial x^i}.$$

Under a coordinate transformation,

$$\frac{\partial b'^i}{\partial x'^i} = \frac{\partial}{\partial x'^i}\left(\frac{\partial x'^i}{\partial x^k}b^k\right) = \frac{\partial x^q}{\partial x'^i}\frac{\partial^2 x'^i}{\partial x^q \partial x^k}b^k + \frac{\partial x^m}{\partial x'^i}\frac{\partial x'^i}{\partial x^k}\frac{\partial b^k}{\partial x^m}$$

$$= b^k\frac{\partial}{\partial x^k}\left(\ln\left|\frac{\partial x'}{\partial x}\right|\right) + \frac{\partial b^k}{\partial x^k}. \tag{4.2.11}$$

The first of these two terms must be counterbalanced in some way if we are to obtain an invariant. Now the metric determinant a transforms so that

$$\sqrt{a'} = \left|\frac{\partial x}{\partial x'}\right|\sqrt{a};$$

thus

$$b'^k\frac{\partial}{\partial x'^k}(\ln\sqrt{a'}) = b^k\frac{\partial}{\partial x^k}\left(\ln\sqrt{a} + \ln\left|\frac{\partial x}{\partial x'}\right|\right). \tag{4.2.12}$$

If we add (4.2.11) and (4.2.12), and note that the Jacobians are reciprocals, we see that the terms containing them cancel. We then find the statement that the scalar

$$\frac{\partial b^i}{\partial x^i} + b^i\frac{\partial}{\partial x^i}\ln\sqrt{a} = \frac{1}{\sqrt{a}}\frac{\partial}{\partial x^i}(\sqrt{a}\,b^i) \tag{4.2.13}$$

is an invariant. This invariant, which fulfils our requirements, is known as the divergence of the vector $\mathbf{b}$, and is denoted symbolically by $\nabla\cdot\mathbf{b}$. Note that the contravariant components of $\mathbf{b}$ appear in the expression for the divergence.

We are now able to define the most important of our invariant differential operators, the Laplacian. As in Cartesian vector calculus, this is a second-order operator, the divergence of the gradient, and it applies to scalar invariants. Thus we have

$$\Delta u \equiv \nabla\cdot\nabla u = \frac{1}{\sqrt{a}}\frac{\partial}{\partial x^i}\left(\sqrt{a}\,a^{ik}\frac{\partial u}{\partial x^k}\right) = a^{ik}\frac{\partial^2 u}{\partial x^i \partial x^k} + \frac{1}{\sqrt{a}}\frac{\partial}{\partial x^i}(\sqrt{a}\,a^{ik})\frac{\partial u}{\partial x^k}. \tag{4.2.14}$$

Clearly the Laplacian Δu is an invariant, since it is the divergence of a vector. The contravariant components a^{ik} appear since it is necessary to raise the covariant index of the partial derivatives before taking the divergence as in (4.2.13). In Δu the coefficients of the second-order derivatives of u are just the original coefficients a^{ik} which appear in $L[u]$.

In fact the operator $L[u]$ may now be written in terms of the Laplacian, and the gradient, of the argument function u. Let us choose a particular coordinate system x^i; then, since the Laplacian Δu encompasses the second-order derivatives and also certain terms with first derivatives of u, we have

$$L[u] = \Delta u + b^i \frac{\partial u}{\partial x^i} + cu,$$

where the b^i are N new functions, and the coefficient c is as in (4.1.1). By definition $L[u]$ is an invariant under coordinate transformations for every scalar invariant u. So also are the terms Δu and cu. Thus the second term on the right must be an invariant. But, by the quotient law of tensor calculus, this implies that the functions b^i are components of a contravariant vector (§ 1.4, Ex. 4). Denoting this vector by the symbol $\mathbf{b}$, we have the expression

$$L[u] = \Delta u + \mathbf{b} \cdot \nabla u + cu, \qquad (4.2.15)$$

which shows that $L[u]$ can be expressed in terms of (1) the Laplacian associated with the fundamental tensor a^{ik}, (2) a vector $\mathbf{b}$, (3) the invariant coefficient c, in addition to u and its derivatives. Each of these three tensorial quantities plays a role in the behaviour of the differential equations connected with this operator $L[u]$.

Exercise 1. Verify that the three expressions for the squared length $\mathbf{b}^2$ in (4.2.4) are equal.

Exercise 2. Calculate the volume of the Riemann space of two dimensions with line element

$$ds^2 = d\theta^2 + \sin^2\theta \, d\varphi^2,$$

where $\qquad 0 \leqslant \theta \leqslant \pi, \qquad 0 \leqslant \varphi \leqslant 2\pi.$

What closed surface in Euclidean three-space has this metric ?

Exercise 3. Calculate the volume of the region $0 \leqslant r \leqslant 1$, $0 \leqslant \theta \leqslant \pi$, $0 \leqslant \varphi \leqslant 2\pi$ in the V_3 with line element

$$ds^2 = dr^2 + R^2 \sinh^2(r/R)[d\theta^2 + \sin^2\theta \, d\varphi^2].$$

Exercise 4. Find the expression for the divergence of a contravariant vector in spherical polar coordinates in Euclidean three-space. Find the expression for the Laplacian in this coordinate system.

Exercise 5. Show that $\nabla \cdot (\mathbf{b}u) = u\nabla \cdot \mathbf{b} + \mathbf{b} \cdot \nabla u$, where u is a scalar invariant and $\mathbf{b}$ a vector.

Exercise 6. Show that $\Delta(uv) = u \Delta v + 2\nabla u \cdot \nabla v + v \Delta u$, for any two scalar invariants u and v. Calculate $\Delta(u^n)$ given Δu and $(\nabla u)^2$.

4.3. Green's formula. The main analytical formula for the treatment of linear second-order equations is Green's formula, which we shall now derive. The first step is to set up the Gauss divergence formula, which expresses the volume integral of a divergence as a surface integral. Let D be a domain or region of V_N, bounded by a smooth (i.e. sufficiently differentiable) surface B, which is an $N-1$ dimensional closed manifold. We shall suppose that B is parametrized by $N-1$ independent parameters t^α ($\alpha = 1,..., N-1$). Let the equation of B have the form

$$\varphi(x^j) = 0. \tag{4.3.1}$$

On B, $x^i = x^i(t^\alpha)$, and therefore

$$\frac{\partial \varphi}{\partial x^j} \frac{\partial x^j}{\partial t^\alpha} = 0. \tag{4.3.2}$$

Equations (4.3.2) may be regarded as $N-1$ linear conditions upon the N quantities $\partial\varphi/\partial x^j$. Since the rank of the matrix $\left(\left(\dfrac{\partial x^j}{\partial t^\alpha}\right)\right)$ is assumed to have its full value $N-1$ on S, the mutual ratios of the partials $\partial\varphi/\partial x^j$ are fully determined by (4.3.2).

Let us form the integral over D of the divergence of a given vector $\mathbf{b}$. We have

$$\int_D \nabla \cdot \mathbf{b} \, dV = \int_D \frac{1}{\sqrt{a}} \frac{\partial}{\partial x^i} (\sqrt{a}\, b^i) \sqrt{a}\, dx^1 \ldots dx^N = \int_D \frac{\partial}{\partial x^i} (\sqrt{a}\, b^i) \, dx^1 \ldots dx^N. \tag{4.3.3}$$

Now let us integrate the differentiated terms over the range of the appropriate variable for each one. We obtain an integral over the boundary surface B: for the kth term this is (k not summed)

$$\int_B \sqrt{a}\, b^k \, dx^1 \ldots dx^{k-1} \, dx^{k+1} \ldots dx^N = \int_B \sqrt{a}\, b^k D_k \, dt^1 \ldots dt^{N-1}, \tag{4.3.4}$$

where
$$D_k = (-1)^{k-1} \frac{\partial(x^1,..., x^{k-1}, x^{k+1},..., x^N)}{\partial(t^1,..., t^{N-1})}. \tag{4.3.5}$$

The N determinants D_k satisfy the $N-1$ linear conditions

$$D_k \frac{\partial x^k}{\partial t^\alpha} = 0, \tag{4.3.6}$$

which can be established by noting that the left-hand side can be written as an $N \times N$ determinant which has two rows the same and therefore vanishes. In view of (4.3.2), conditions (4.3.6) imply that the $\partial\varphi/\partial x^k$ and D_k are proportional:

$$\frac{\partial \varphi}{\partial x^k} = \alpha D_k \tag{4.3.7}$$

for some scalar factor of proportionality α. Since the $\partial\varphi/\partial x^k$ determine the unit normal to B, namely,

$$n_k = \frac{1}{\sqrt{\{(\nabla\varphi)^2\}}}\frac{\partial\varphi}{\partial x^k},$$

it follows that
$$n_k = \frac{D_k}{\sqrt{(a^{ij}D_i D_j)}} = \frac{D_k}{D}. \qquad (4.3.8)$$

Here
$$D^2 = a^{ij}D_i D_j = a^{ij}\frac{\partial\varphi}{\partial x^i}\frac{\partial\varphi}{\partial x^j}\alpha^{-2} \qquad (4.3.9)$$

is different from zero provided that the surface B is not a characteristic surface at the point in question.

From (4.3.3), (4.3.4), and (4.3.8) we see that

$$\int_D \nabla\cdot\mathbf{b}\,dV = \int_B b^k D_k\,\sqrt{a}\,dt^1\dots dt^{N-1}$$

$$= \int_B b^k n_k\,\sqrt{a}\,D\,dt^1\dots dt^{N-1} \qquad (4.3.10)$$

$$= \int_B \mathbf{b}\cdot\mathbf{n}\,dS,$$

where
$$dS = \sqrt{a}\,D\,dt^1\dots dt^{N-1} \qquad (4.3.11)$$

is the element of surface area on B. This is the Gauss divergence formula.

Exercise 1. In Euclidean three-space, show that, according to (4.3.11),
$$dS = \sqrt{(EG - F^2)}\,du\,dv$$
in the notation of differential geometry (12a, vol. i).

Exercise 2. If s is arc length along the normal to the surface, show that $dV = ds\,dS$. *Hint*: Let s, t^α be taken as coordinates.

In order to apply formula (4.3.10), we note the following identities, which are either already proved or are easy consequences of previous results. Let u and v be two suitably differentiable but otherwise arbitrary functions. Then
$$u\,\Delta v + \nabla u\cdot\nabla v = \nabla\cdot(u\nabla v) \qquad (4.3.12)$$
and
$$\left.\begin{array}{c} u\,\Delta v - v\,\Delta u = \nabla\cdot(u\nabla v - v\nabla u) \\ v\mathbf{b}\cdot\nabla u + u\nabla\cdot(\mathbf{b}v) = \nabla\cdot(\mathbf{b}uv) \end{array}\right\}, \qquad (4.3.13)$$

where in (4.3.13), $\mathbf{b}$ is a vector. Therefore
$$v[\Delta u + \mathbf{b}\cdot\nabla u + cu] - u[\Delta v - \nabla\cdot(\mathbf{b}v) + cv]$$
$$= \nabla\cdot(v\nabla u - u\nabla v + \mathbf{b}uv). \qquad (4.3.14)$$

Formula (4.3.14) may be written

$$vL[u] - uM[v] = \nabla \cdot (v\nabla u - u\nabla v + \mathbf{b}uv), \qquad (4.3.15)$$

where

$$M[v] = \Delta v - \nabla \cdot (\mathbf{b}v) + cv \qquad (4.3.16)$$

is a second linear operator, known as the adjoint of $L[u]$. The defining property of the adjoint operator is formula (4.3.15), in which the difference on the left-hand side is expressed as a divergence.

From (4.3.10) and (4.3.15) we see that

$$\int_D \{vL[u] - uM[v]\}\, dV = \int_D \nabla \cdot (u\nabla v - v\nabla u + \mathbf{b}uv)\, dV$$

$$= \int_B (v\nabla u - u\nabla v + \mathbf{b}uv) \cdot \mathbf{n}\, dS \qquad (4.3.17)$$

$$= \int_B \left(v\frac{\partial u}{\partial n} - u\frac{\partial v}{\partial n} + b_n\, uv \right) dS,$$

where, by definition (4.2.9), the normal derivative indicates the scalar product of the unit normal $\mathbf{n}$ with the gradient ∇u:

$$\frac{\partial u}{\partial n} = \mathbf{n} \cdot \nabla u = a^{ik} n_i \frac{\partial u}{\partial x^k}. \qquad (4.3.18)$$

Formula (4.3.17) is Green's second formula; thus

$$\int_D \{vL[u] - uM[v]\}\, dV = \int_B \left\{ v\frac{\partial u}{\partial n} - u\frac{\partial v}{\partial n} + b_n\, uv \right\} dS. \qquad (4.3.19)$$

Here b_n denotes the normal component $\mathbf{b} \cdot \mathbf{n}$ of the vector $\mathbf{b}$. It is assumed that the bounding surface B of D contains no regions which are characteristic surfaces.

In an important special case we can obtain a different version of Green's formula. The operator $L[u]$ is said to be self-adjoint if its adjoint $M[v]$ coincides with $L[v]$. From (4.3.16) we see that this is the case if and only if

$$\mathbf{b} \cdot \nabla u = -\nabla \cdot (\mathbf{b}u) = -\mathbf{b} \cdot \nabla u - u\nabla \cdot \mathbf{b}. \qquad (4.3.20)$$

Setting $u = 1$ in (4.3.20), we see that $\nabla \cdot \mathbf{b} = 0$, and, setting $u = x^i$ ($i = 1,...,N$), it then follows that $b^i = 0$ ($i = 1,...,N$), so $b \equiv 0$. Thus $L[u]$ is self-adjoint if and only if $b = 0$. Every self-adjoint operator has therefore the form

$$L[u] = \Delta u + cu. \qquad (4.3.21)$$

Suppose $L[u]$ is self-adjoint. From equation (4.3.12) we have

$$v(\Delta u + cu) + \nabla u \cdot \nabla v - cuv = \nabla \cdot (v\nabla u), \qquad \text{whence the}$$

Gauss formula yields, as in (4.3.17),

$$\int_D vL[u]\, dV + \int_D \{\nabla u \cdot \nabla v - cuv\}\, dV = \int_B v\frac{\partial u}{\partial n}\, dS. \qquad (4.3.22)$$

Historically, this was Green's first formula. The integral

$$E(u,v) = \int_D \{\nabla u \cdot \nabla v - cuv\}\, dV \qquad (4.3.23)$$

is known as the Dirichlet integral of the operator L relative to the domain D. If $L[u]$ is elliptic, and if c is negative, then the Dirichlet integral is positive definite in the sense that

$$E(u,u) = \int_D \{(\nabla u)^2 - cu^2\}\, dV > 0 \qquad (4.3.24)$$

for all once differentiable functions u not identically zero in D. This property of the Dirichlet integral will be important in the theory of elliptic equations.

Exercise 3. Show that

$$E(\lambda u + \mu v, \lambda u + \mu v) = \lambda^2 E(u,u) + 2\lambda\mu E(u,v) + \mu^2 E(v,v)$$

if λ and μ are constants.

Exercise 4. Show that if $c \leqslant 0$, and if $L[u]$ is elliptic,

$$E(u,v)^2 \leqslant E(u,u)E(v,v).$$

Exercise 5. Show that if u is twice continuously differentiable in D and on B, then

$$\int_B \frac{\partial u}{\partial n}\, dS = \int_D \Delta u\, dV.$$

4.4. Flat space. Equations with constant coefficients. A space V_N provided with a Riemann metric is said to be flat if there exists a co-ordinate system in which the components of the fundamental tensor are constants. Otherwise the space is 'curved'. If V_N is flat, it is possible to bring the line element (4.2.5) to normal form valid at every point of the space, and we may assume that this has been done. The tensor calculus also provides an explicit though lengthy test to determine whether or not a space is flat. The criterion is that the Riemann curvature tensor must vanish identically. We shall not pursue this question here but refer the reader to (32, ch. iii).

For the differential operator $L[u]$, the flatness of the Riemann space implies that there is a coordinate system such that the coefficients a^{ik}

of the second-order derivatives are all constants. Thus reduction to normal form for the line element applies also to the operator $L[u]$. If, in this normal coordinate system, the coefficients of the other terms in $L[u]$ are also constants, we see that $L[u]$ is an operator with constant coefficients. The equation $L[u] = 0$ has in this case especially simple properties which we shall investigate in the next few sections. The two most important special cases are:

(a) $L[u]$ elliptic:

$$L[u] = \frac{\partial^2 u}{(\partial x^1)^2} + \frac{\partial^2 u}{(\partial x^2)^2} + \cdots + \frac{\partial^2 u}{(\partial x^N)^2} + b^i \frac{\partial u}{\partial x^i} + cu = 0.$$

(b) $L[u]$ normal hyperbolic (Lorentzian):

$$L[u] = \frac{\partial^2 u}{(\partial x^1)^2} - \frac{\partial^2 u}{(\partial x^2)^2} - \cdots - \frac{\partial^2 u}{(\partial x^N)^2} + b^i \frac{\partial u}{\partial x^i} + cu = 0.$$

If the vector b^i has constant components, we can, by a change of the dependent variable u, eliminate the terms involving first derivatives. We shall do this for case (a), leaving (b) as an exercise to the reader. For the elliptic operator $L[u]$, let us set

$$v = u \exp\left[\tfrac{1}{2} \sum_i b^i x^i\right],$$

and find the differential equation for v. We have

$$\frac{\partial v}{\partial x^j} = \exp\left[\tfrac{1}{2} \sum_i b^i x^i\right]\left\{\frac{\partial u}{\partial x^j} + \tfrac{1}{2} b^j u\right\},$$

$$\frac{\partial^2 v}{(\partial x^j)^2} = \exp\left[\tfrac{1}{2} \sum_i b^i x^i\right]\left\{\frac{\partial^2 u}{(\partial x^j)^2} + b^j \frac{\partial u}{\partial x^j} + \tfrac{1}{4}(b^j)^2 u\right\}.$$

The equation for v is therefore

$$\frac{\partial^2 v}{(\partial x^1)^2} + \cdots + \frac{\partial^2 v}{(\partial x^N)^2} + \left[c - \tfrac{1}{4} \sum_i (b^i)^2\right]v = 0.$$

This equation has no first derivatives appearing in it.

Exercise. Can a parabolic differential equation of the second order with constant coefficients be simplified in this way?

Linear differential equations with constant coefficients are much easier to discuss and to solve than equations with variable coefficients, because it is possible to find explicit formulae for many particular solutions. Furthermore, these formulae are usually valid over the whole domain of the independent variables. Thus we are not forced to work 'locally' as is so often necessary with equations having variable coefficients.

The technique of integration for linear equations of the second order is based on Green's formula, together with certain special 'fundamental' solutions of the differential equation, and of its adjoint equation. These special solutions are useful because they have a singularity, or infinity, which may be used in conjunction with Green's formula to develop a calculus of residues not unlike that based on Cauchy's integral formula in the theory of functions of a complex variable. We shall develop this technique in a later chapter; we would now like to show that equations with constant coefficients do always possess a fundamental solution defined in the entire space of the independent variables.

Let us show this for the (elliptic) Laplace equation in $N > 2$ dimensions:

$$\Delta u = \frac{\partial^2 u}{(\partial x^1)^2} + \dots + \frac{\partial^2 u}{(\partial x^N)^2} = 0.$$

The Riemannian geometry based on this equation is precisely the geometry of N-dimensional Euclidean space E_N. Let O be the origin of Cartesian coordinates x^i in E_N. Let us find, if possible, a solution which depends only on the distance $r = OP$ where P is the current coordinate point. Since

$$OP^2 = r^2 = (x^1)^2 + \dots + (x^N)^2,$$

we see that

$$\frac{\partial r^a}{\partial x^i} = \tfrac{1}{2} a r^{a-2} \frac{\partial r^2}{\partial x^i} = a r^{a-2} x^i,$$

and

$$\frac{\partial^2 r^a}{(\partial x^i)^2} = a(a-2) r^{a-4} (x^i)^2 + a r^{a-2}.$$

Summing this last formula over the N values of the index i, we find

$$\Delta r^a = a(a-2) r^{a-4} \sum_i (x^i)^2 + a N r^{a-2}$$

$$= a(a-2+N) r^{a-2}.$$

Thus, if $a = 0$, or $a = -N+2$, r^a is a solution of the differential equation. The choice $a = 0$ leads to a trivial solution. However, the other solution r^{-N+2} is the fundamental solution (often called a fundamental singularity) for this differential equation. Note that the solution has indeed an infinity of order $N-2$ at the origin. The order of this infinity is uniquely determined by the requirement that the solution should be spherically symmetric about the origin, that is, should depend only upon the distance OP. As base point we may take any point Q, in which case

$$r^2 = r^2_{PQ} = \sum_i (x^i - y^i)^2,$$

where x^i are the coordinates of P, y^i the coordinates of Q. To sum up, then, the function

$$\gamma(P, Q) = r^{-N+2} \tag{4.4.1}$$

is a fundamental solution with argument point P, and parameter point Q.

Exercise 1. The argument and parameter points may be interchanged, in other words, r^{-N+2} is a solution as a function of Q.

Exercise 2. If $N = 2$, this solution is a constant. Show that $\log r$ is then a solution.

The most general elliptic equation with constant coefficients, which, as we have seen, may be written

$$\Delta u + cu = 0 \quad (c = \text{constant}),$$

where Δ has its previous meaning, has also a radial fundamental solution with a singularity of order $N-2$. Let $\gamma = \gamma(r)$ be any radial solution, then, by a calculation in partial derivatives which we leave the reader to verify, we see that $\gamma(r)$ satisfies the ordinary differential equation:

$$\gamma'' + \frac{N-1}{r}\gamma' + c\gamma = 0.$$

If we introduce as dependent variable a new function $v(\rho)$, where $\rho = \sqrt{c}\,r$, defined by the relation

$$\gamma(r) = r^{-\frac{1}{2}(N-2)}v(\sqrt{c}\,r),$$

then $v(\rho)$ satisfies the transformed ordinary differential equation

$$v''(\rho) + \frac{1}{\rho}v'(\rho) + \left[1 - \frac{(N-2)^2}{4\rho^2}\right]v = 0.$$

This is Bessel's equation of order $\frac{1}{2}(N-2)$, and it possesses a solution which becomes infinite for $\rho = 0$. We may define the fundamental solution as the function $\gamma(r)$ which corresponds to this solution of the Bessel equation. In the usual notation for Bessel functions, we have for N odd,

$$\gamma(r) = r^{-\frac{1}{2}(N-2)}J_{-\frac{1}{2}(N-2)}(\sqrt{c}\,r), \tag{4.4.2}$$

while for even N, $\gamma(r) = r^{-\frac{1}{2}(N-2)}N_{\frac{1}{2}(N-2)}(\sqrt{c}\,r).$ $\tag{4.4.3}$

Here $N_p(x)$ denotes the Neumann function of order p. It may be verified that the order of the singularity for $r = 0$ is $N-2$, in either case, unless $N = 2$, in which case $N_0(x)$ has a logarithmic infinity.

Exercise. For the hyperbolic equation $\Delta u = 0$, of type (b), p. 85, find a solution of the form r^a, where $r^2 = (x^1 - y^1)^2 - (x^2 - y^2)^2 - \ldots - (x^N - y^N)^2$. Discuss the locus of singularities of this solution.

The elliptic Laplace equation with constant coefficients has a striking 'mean-value' property. Let P be an arbitrary point, r a variable distance, and let us denote the element of solid angle at P by $d\Omega$. Then the average value of $u(R)$ on the surface S of the sphere K with centre P, radius r,

$$v(P,r) = \frac{1}{\omega_N} \int u \, d\Omega, \quad \text{where } \omega_N = \int d\Omega, \qquad (4.4.4)$$

is equal to the value of $u(P)$. To show this, we calculate the derivative of $v(P,r)$ with respect to r, as follows:

$$\frac{dv}{dr} = \frac{1}{\omega_N} \int \frac{\partial u}{\partial r} d\Omega = \frac{1}{\omega_N r^{N-1}} \int \frac{\partial u}{\partial r} r^{N-1} d\Omega$$

$$= \frac{1}{\omega_N r^{N-1}} \int_S \frac{\partial u}{\partial n} dS = \frac{1}{\omega_N r^{N-1}} \int_K \Delta u \, dV = 0,$$

using Exercise 3 of § 4.3, and the fact that $\Delta u = 0$. Thus $v(P,r)$ is independent of r; we see at once that $v(P,0) = u(P)$, and this proves $v(P,r) = u(P)$. *A harmonic function in Euclidean space is equal to its mean value over any concentric sphere.*

Exercise. Any maximum or minimum values of such a harmonic function $u(P)$ are attained on the boundary of the region in which $u(P)$ is harmonic.

4.5. Geodesics and geodesic distance. As we have seen, the fundamental solution of the Laplace equation in a flat space can be expressed as a function of the Euclidean distance $r(P, Q)$ between the two argument points. In a curved Riemann space, where the finite distance measurements of Euclidean space break down, we shall need to find some generalization of the separation, or distance, between two fixed points, in order to construct a fundamental solution. Such a generalization is available to us in the theory of geodesics. The straight lines of Euclidean space enjoy the property of having the shortest Euclidean arc length of any curve joining two points of the line; and this minimum property will serve to define geodesic lines in a curved space as well. First we shall find the differential equations which characterize geodesic lines, and then we shall derive certain useful properties of the geodesic distance $s(P, Q)$ defined as the arc length along the geodesic line from P to Q. For simplicity, let us consider a positive definite metric. The corresponding results for an indefinite metric are valid if attention is paid to the choice of sign in some of the equations, except for the geodesics of zero length which we shall study in a later chapter.

Let P and Q be two given points, and let a family of smooth curves joining P to Q be drawn. If v be used to label the curves of the family, and u $(0 \leqslant u \leqslant 1)$ is a parameter along each curve, the equations of the curves may be written

$$x^r = x^r(u, v) \tag{4.5.1}$$

with $\qquad x^r(0, v) = x^r(P); \qquad x^r(1, v) = x^r(Q).$

For each curve we have an arc length, which depends on the curve, that is, on v. Denote this arc length by $F(v)$; thus

$$F(v) = \int_P^Q ds = \int_0^1 (a_{rs} p^r p^s)^{\frac{1}{2}} \, du = \int_0^1 w^{\frac{1}{2}} \, du, \tag{4.5.2}$$

where $\qquad p^r = \dfrac{dx^r(u, v)}{du}; \qquad w = a_{rs} p^r p^s. \tag{4.5.3}$

If the arc length is to be minimal (or stationary) we must have $F'(v) = 0$. Now

$$F'(v) = \int_0^1 \frac{d}{dv}(w^{\frac{1}{2}}) \, du = \int_0^1 \left\{ \frac{dx^r}{dv} \frac{\partial w^{\frac{1}{2}}}{\partial x^r} + \frac{dp^r}{dv} \frac{\partial w^{\frac{1}{2}}}{\partial p^r} \right\} du, \tag{4.5.4}$$

since w depends on the x^r and also upon the p^r. Now we have

$$\frac{dp^r}{dv} = \frac{d}{dv} \frac{dx^r}{du} = \frac{d}{du} \frac{dx^r}{dv}, \tag{4.5.5}$$

and also, since the end-points P and Q are fixed,

$$\frac{dx^r}{dv} = 0 \quad \text{(for } u = 0 \text{ or } u = 1 \text{)}. \tag{4.5.6}$$

The second term in the integrand of (4.5.4) now becomes

$$\frac{d}{du}\left(\frac{dx^r}{dv}\right) \cdot \frac{\partial w^{\frac{1}{2}}}{\partial p^r},$$

so we may integrate by parts; and, according to (4.5.6), the integrated terms will fall out. Thus, finally,

$$F'(v) = \int_0^1 \left\{ \frac{\partial w^{\frac{1}{2}}}{\partial x^r} - \frac{d}{du}\left(\frac{\partial w^{\frac{1}{2}}}{\partial p^r}\right) \right\} \frac{dx^r}{dv} \, du. \tag{4.5.7}$$

Suppose now that we have found a shortest curve joining P and Q. This curve may be imbedded in a family of curves in which, at any point, the dx^r/dv can be chosen independently and arbitrarily. But, for all such families, $F'(v)$ is necessarily zero. This can only happen if the coefficient of dx^r/dv in the integrand vanishes identically for $r = 1,..., N$. Thus we see that the Eulerian differential equations

$$\frac{d}{du}\left(\frac{\partial w^{\frac{1}{2}}}{\partial p^r}\right) - \frac{\partial w^{\frac{1}{2}}}{\partial x^r} = 0 \qquad (4.5.8)$$

are satisfied at every point of the shortest curve. We shall broaden our view-point slightly, and say that any curve which satisfies these equations shall be called a geodesic curve. In view of (4.5.3), equations (4.5.8) are N ordinary differential equations of the second order, in the N dependent variables x^r.

These equations can be written in a more explicit form if we select s, the arc length, as independent variable. With this choice, we see that w, defined by (4.5.3), is just the square length of the unit tangent vector dx^r/ds to the curve, and is therefore equal to unity. Then we have

$$\frac{d}{ds}\left(\frac{\partial w^{\frac{1}{2}}}{\partial p^r}\right) = \frac{d}{ds}\left(\frac{w^{-\frac{1}{2}}}{2}\frac{\partial w}{\partial p^r}\right) = \frac{d}{ds}(a_{mr}p^m)$$

$$= a_{rm}\frac{dp^m}{ds} + \frac{\partial}{\partial x^n}(a_{mr})\frac{dx^n}{ds}p^m$$

$$= a_{rm}\frac{dp^m}{ds} + \frac{1}{2}\left\{\frac{\partial a_{mr}}{\partial x^n} + \frac{\partial a_{nr}}{\partial x^m}\right\}p^m p^n. \qquad (4.5.9)$$

Furthermore, $\qquad \dfrac{\partial}{\partial x^r}w^{\frac{1}{2}} = \frac{1}{2}w^{-\frac{1}{2}}\dfrac{\partial w}{\partial x^r} = \dfrac{1}{2}\dfrac{\partial a_{mn}}{\partial x^r}p^m p^n. \qquad (4.5.10)$

It now follows that the equations (4.5.8) may be written

$$a_{rm}\frac{d^2x^m}{ds^2} + \frac{1}{2}\left(\frac{\partial a_{mr}}{\partial x^n} + \frac{\partial a_{nr}}{\partial x^m} - \frac{\partial a_{mn}}{\partial x^r}\right)p^m p^n = 0. \qquad (4.5.11)$$

Since the determinant of the coefficients a_{mr} of the second derivatives is not zero, we may solve for those derivatives. The final form of the equations is usually written

$$\frac{d^2x^r}{ds^2} + \left\{\begin{matrix} r \\ mn \end{matrix}\right\}\frac{dx^m}{ds}\frac{dx^n}{ds} = 0, \qquad (4.5.12)$$

where the expression (it is not a tensor)

$$\left\{\begin{matrix} r \\ mn \end{matrix}\right\} = \frac{a^{rs}}{2}\left(\frac{\partial a_{sm}}{\partial x^n} + \frac{\partial a_{sn}}{\partial x^m} - \frac{\partial a_{mn}}{\partial x^s}\right) \qquad (4.5.13)$$

is known as a Christoffel symbol of the second kind. Note that the Christoffel symbols vanish if the a_{rs} are all constants. The reader will verify easily that the geodesic lines are, in this latter case, 'straight lines'. In general, a geodesic line will be uniquely determined by initial values of the x^r and

$p^r = dx^r/ds$; that is, by a point and a direction, just as a straight line is determined in Euclidean space. When two given points are sufficiently close, a well-determined geodesic line can be drawn joining them.

Exercise. What curves are the geodesics on a sphere? Show that two points of the sphere can always be joined by two geodesics.

We now define $s = s(P, Q) = s(Q, P)$, a scalar invariant function of two points P and Q, as the arc length of the shortest geodesic line joining P and Q. Also we shall set

$$\Gamma(P, Q) = s^2(P, Q). \tag{4.5.14}$$

With an indefinite metric it is often convenient to use Γ instead of s, thus avoiding imaginary quantities, because s, being the integral of ds, will be imaginary if $ds^2 < 0$.

Exercise. Find the formula for $\Gamma(P, Q)$ in terms of the coordinates of P and Q in a flat space with Cartesian coordinates.

Suppose now that Q is fixed, and consider $s(P, Q)$ as a function of P. We shall need to evaluate the square length of the gradient vector ∇s. Clearly s increases most rapidly when P varies along the geodesic from Q through the initial position of P. That is, $\nabla_P s(P, Q)$ has the direction of the tangent vector at P to the geodesic from Q to P. Furthermore, the change of s in a displacement of length ds along this geodesic is precisely ds. Thus the magnitude of ∇s is unity:

$$(\nabla s)^2 = a_{rm} \frac{\partial s}{\partial x^r} \frac{\partial s}{\partial x^m} = 1. \tag{4.5.15}$$

Since

$$\nabla \Gamma = \nabla(s^2) = 2s\nabla s, \tag{4.5.16}$$

we have

$$(\nabla \Gamma)^2 = 4s^2(\nabla s)^2 = 4\Gamma. \tag{4.5.17}$$

In these two formulae, either argument point may be taken as the one on which ∇ operates. Equation (4.5.17) will be important in the study of the characteristics of hyperbolic equations, to which we shall return in Chapter IX.

Again, let $U(P)$ be any scalar invariant, and let s denote arc length along the geodesic through P in a given direction. The directional derivative dU/ds is by definition the rate of change of U with respect to s along this curve. But dU/ds is also the scalar product of ∇U with the unit vector tangent to the geodesic. In view of (4.5.15), therefore, we have

$$\frac{dU}{ds} = \nabla U \cdot \nabla s. \qquad (4.5.18)$$

Multiplying (4.5.18) by $2s$, we have, from (4.5.16),

$$2s\frac{dU}{ds} = \nabla U \cdot \nabla \Gamma. \qquad (4.5.19)$$

Exercise. If $F(\Gamma)$ is a function of Γ alone, then

$$\nabla F(\Gamma) \cdot \nabla U = 2F'(\Gamma)s\frac{dU}{ds}.$$

In addition to the formulae just derived, we shall later need expressions for the Laplacians of certain quantities connected with Γ. First, for any function $F(\Gamma)$ of Γ alone, we have, in some coordinate system,

$$
\begin{aligned}
\Delta F(\Gamma) &= \frac{1}{\sqrt{a}}\frac{\partial}{\partial x^r}\left(\sqrt{a}\,a^{rs}\frac{\partial F(\Gamma)}{\partial x^s}\right) = \frac{1}{\sqrt{a}}\frac{\partial}{\partial x^r}\left(F'(\Gamma)\sqrt{a}\,a^{rs}\frac{\partial\Gamma}{\partial x^s}\right) \\
&= F'(\Gamma)\Delta\Gamma + a^{rs}\frac{\partial\Gamma}{\partial x^s}\frac{\partial}{\partial x^r}F'(\Gamma) \\
&= F'(\Gamma)\Delta\Gamma + F''(\Gamma)a^{rs}\frac{\partial\Gamma}{\partial x^s}\frac{\partial\Gamma}{\partial x^r} \\
&= F'(\Gamma)\Delta\Gamma + 4F''(\Gamma).\Gamma,
\end{aligned}
\qquad (4.5.20)
$$

the last step being a consequence of (4.5.17). The formula for the Laplacian of a product UV follows easily from the definition of the Laplacian as the divergence of the gradient, and is

$$\Delta(UV) = U\Delta V + 2\nabla U \cdot \nabla V + V\Delta U. \qquad (4.5.21)$$

Using (4.5.19), (4.5.20), and (4.5.21), we see that

$$
\begin{aligned}
\Delta\{F(\Gamma)U\} &= U\Delta F(\Gamma) + 2F'(\Gamma)\nabla\Gamma \cdot \nabla U + F(\Gamma)\Delta U \\
&= U[F'(\Gamma)\Delta\Gamma + 4F''(\Gamma)\Gamma] + 4F'(\Gamma)s\frac{dU}{ds} + F(\Gamma)\Delta U. \quad (4.5.22)
\end{aligned}
$$

In particular, if $F(\Gamma) = \Gamma^\beta$, we have

$$\Delta(\Gamma^\beta U) = \beta\Gamma^{\beta-1}\left[\{4(\beta-1)+\Delta\Gamma\}U + 4s\frac{dU}{ds}\right] + \Gamma^\beta\Delta U. \qquad (4.5.23)$$

This formula will be applied in the construction of the fundamental singularity for elliptic equations, and also in the construction of the Riesz kernel in Chapter X.

In order to simplify certain expressions which involve Γ we shall need to introduce special coordinate systems which have many properties in common with Cartesian coordinates, provided that the origin is held fixed. These are known as Riemannian or normal coordinates, and are so chosen that the geodesic lines through the 'pole', or origin, appear as straight lines. To set up a Riemannian coordinate system, we select a point P as pole. Let $\{X_0^k\}$ be a coordinate system, and consider any geodesic line through P. This geodesic has a unit tangent vector at P, which we shall denote by

$$\dot{x}_0^k = \frac{dx_0^k}{ds} = p_0^k. \qquad (4.5.24)$$

Here
$$a_{ik}(P)p_0^i p_0^k = 1. \qquad (4.5.25)$$

Now each point Q sufficiently close to P can be joined to P by a unique shortest geodesic line, the geodesic distance being $s(P, Q)$. We define the Riemannian coordinates of Q to be

$$x^k = p_0^k s(P, Q) = p_0^k s. \qquad (4.5.26)$$

From (4.5.26) we see that, along a fixed geodesic line through P, s varies but the p_0^k remain fixed. Hence, if p^k denotes the unit tangent vector at a point on this geodesic, we have

$$p^k = \frac{dx^k}{ds} = p_0^k. \qquad (4.5.27)$$

Therefore $p^k = $ constant along the geodesic, so that, in the Riemannian coordinate system, the tangent vector has the same direction at each point of the geodesic. In other words, the geodesic is a 'straight line' through the origin. This fact also emerges if we regard (4.5.26) as linear parametric equations with parameter s.

In view of (4.5.27), the second derivative $\dfrac{d^2x^k}{ds^2} = \dfrac{dp^k}{ds}$ is zero. Referring back to (4.5.12), the equations of the geodesics, we see that

$$\begin{Bmatrix} r \\ mn \end{Bmatrix} p^m p^n = 0,$$

or, multiplying by s^2 and using $p^k = p_0^k$, that

$$\begin{Bmatrix} r \\ mn \end{Bmatrix} x^m x^n = 0. \qquad (4.5.28)$$

If a coordinate system is Riemannian with pole at P, equations (4.5.28) hold.

Exercise 1. If (4.5.28) hold, show that $\{x^k\}$ is Riemannian.

Exercise 2. In Riemannian coordinates, show that

$$x^r \frac{\partial U}{\partial x^r} = s \frac{dU}{ds}.$$

Riemannian coordinates enable us to give a very simple representation for the square $\Gamma(P, Q)$ of the geodesic distance from the pole P. From (4.5.25) we see that

$$\Gamma(P, Q) = s^2(P, Q)$$

$$= a_{mn}(P) p_0^m p_0^n s^2(P, Q)$$

$$= a_{mn}(P) x^m x^n, \tag{4.5.29}$$

the last step being a consequence of (4.5.26). Thus Γ is a quadratic form in Riemannian coordinates. It must be remembered that if P varies, the entire coordinate system in which (4.5.29) is valid will change. In a flat space, however, where Riemannian coordinates become Cartesian co-ordinates, the representation (4.5.29) is valid when both P and Q vary, provided that instead of x^m we supply the difference $x^m(P) - x^m(Q)$ of the coordinates of P and Q in a Cartesian coordinate system.

Exercise 1. Show that $\Gamma(P, Q) = a_{mn}(R) x^m x^n$, where R is any point on the geodesic from P to Q.

Exercise 2. Show that there exists a coordinate system with origin at P such that

$$\Gamma(P, Q) = \sum_i (x^i)^2$$

if the metric is positive definite.

The quantity $\Delta\Gamma$ may also be expressed rather simply in Riemannian coordinates. Let us suppose that Δ is taken with respect to the argument point P, and that P is the pole of the Riemannian coordinates. Then we have

$$\Delta\Gamma(P, Q) = \frac{1}{\sqrt{a}} \frac{\partial}{\partial x^r} \left(\sqrt{a}\, a^{rs} \frac{\partial \Gamma}{\partial x^s} \right) = \frac{1}{\sqrt{a}} \frac{\partial}{\partial x^r} (\sqrt{a}\, a^{rs}\, 2 a_{st}\, x^t), \tag{4.5.30}$$

in view of Exercise 1 above.

Since $a^{rs} a_{st} = \delta_t^r$, we see that

$$\Delta\Gamma = \frac{2}{\sqrt{a}} \frac{\partial}{\partial x^r} (\sqrt{a}\, x^r) = 2 \left[\frac{\partial x^r}{\partial x^r} + \frac{x^r}{\sqrt{a}} \frac{\partial}{\partial x^r} \sqrt{a} \right]$$

$$= 2N + 2x^r \frac{\partial}{\partial x^r} \ln \sqrt{a}$$

$$= 2N + s \frac{d \ln \sqrt{a}}{ds}. \tag{4.5.31}$$

The evaluation of $\partial x^r/\partial x^r$ is an obvious consequence of the summation convention. We see that $\Delta\Gamma$ depends only on the value of the determinant a which measures the volume element in Riemannian coordinates.

Exercise 1. If the initial coordinate system x_0^i is transformed, show that the Riemannian coordinates x^k of a point Q transform like a contravariant vector at the pole P.

Exercise 2. Show that the first derivatives $\partial a_{mn}/\partial x^r$ of the components of the metric tensor in Riemannian coordinates are zero at the pole P.

Exercise 3. Show that

$$\Delta\Gamma = 2N + O(s^2),$$

in a positive definite metric. Is this still true in an indefinite metric?

Exercise 4. If x^i are Riemannian coordinates with pole at P, and if $a_{ik} = 1 \ (i = k)$, $a_{ik} = 0 \ (i \neq k)$, and if $s, \theta^1,..., \theta^{N-1}$ are defined by

$$x^1 = s \sin\theta^1 \sin\theta^2 ... \sin\theta^{N-2} \sin\theta^{N-1},$$

$$x^2 = s \sin\theta^1 \sin\theta^2 ... \sin\theta^{N-2} \cos\theta^{N-1},$$

$$\vdots$$

$$x^i = s \sin\theta^1 \sin\theta^2 ... \sin\theta^{N-i} \cos\theta^{N-i+1},$$

$$\vdots$$

$$x^N = s \cos\theta^1,$$

show that s is geodesic distance from P, and that $dV = s^{N-1} \, ds \, d\Omega$ where

$$d\Omega = \sin^{N-2}\theta^1 \sin^{N-3}\theta^2 ... \sin\theta^{N-2} \, d\theta^1 ... d\theta^{N-1}$$

is the element of solid angle in Euclidean N-space. The coordinates $s, \theta^1,..., \theta^{N-1}$ are known as geodesic polars.

EXERCISES.   CHAPTER IV

Exercise 1. Show that if $N = 2$, the type of the operator $L[u]$ is determined by the sign of the determinant of the coefficients of the second derivatives.

Exercise 2. Show that

$$u(x^i, t) = t^{-\frac{1}{2}N} \exp\left[-\frac{\sum (x^i)^2}{4t}\right]$$

is a solution of the parabolic equation

$$\frac{\partial u}{\partial t} = \frac{\partial^2 u}{(\partial x^1)^2} + ... + \frac{\partial^2 u}{(\partial x^N)^2}.$$

Hint: Let $r^2 = \sum (x^i)^2$ and note that u is a function of r and t only.

Exercise 3. Show that

$$u(x^i) = \int_a^r \csc(s/R)\, ds$$

is a harmonic function in the V_2 with metric

$$ds^2 = dr^2 + R^2 \sin^2(r/R)\, d\theta^2.$$

Characterize this space geometrically. Show that

$$u(x^i) = \int_a^r \csc^2(s/R)\, ds$$

is a harmonic function in the V_3 with metric

$$ds^2 = dr^2 + R^2 \sin^2(r/R)\, (d\theta^2 + \sin^2\theta\, d\varphi^2).$$

Generalize these results to an N-dimensional Riemannian space which may be represented as a sphere imbedded in a Euclidean space of $N+1$ dimensions. Show that the solutions thus constructed have a singularity of order $N-2$ at the origin in the space, and also at the antipodal point where $r = \pi R$.

Exercise 4. Extend the mean value property of Laplace's equation (§ 4) to the equation with constant coefficients

$$\Delta u + cu = 0,$$

by showing that the mean value $v(P, r)$ is now a Bessel function of $\sqrt{c}\, r$.

Exercise 5. In a Euclidean space E_N, $N > 2$, let $\omega_s(P)$ be the solid angle subtended at a point P by a smooth surface S. Show that

$$\int_S \frac{\partial}{\partial n} [r^{-N+2}]\, dS = -(N-2)\omega_s(P),$$

where $r = r(P, Q)$ and Q ranges over the surface. If S is a closed surface, show that the integral is constant in the interior and zero outside the surface.

Exercise 6. If $L[u]$ and $M[u]$ are defined as in (4.2.15) and (4.3.16), and

$$E(u, v) = \int_D \{\nabla u \cdot \nabla v - [c - \tfrac{1}{2}\nabla \cdot b]uv\}\, dV,$$

show that

$$2E(u, v) + \int_D \{uL[v] + vL[u]\}\, dV = \int_B \left\{u\frac{\partial v}{\partial n} + v\frac{\partial u}{\partial n} + b_n uv\right\} dS,$$

$$2E(u, v) + \int_D \{uM[v] + vM[u]\}\, dV = \int_B \left\{u\frac{\partial v}{\partial n} + v\frac{\partial u}{\partial n} - b_n uv\right\} dS.$$

Exercise 7. If in $L[u] = \Delta u + \mathbf{b} \cdot \nabla u + cu = 0$ a transformation $u = \lambda v$ to a new dependent variable is made, and the equation for v is written out in similar form, show that the invariant

$$c - \tfrac{1}{2}\nabla \cdot \mathbf{b} - \tfrac{1}{4}b^2$$

is unchanged by the transformation.

V

SELF-ADJOINT ELLIPTIC EQUATIONS

THE first equations of elliptic type to be studied in detail were the Laplace equations in Euclidean space of two or of three dimensions. In two dimensions harmonic functions are closely connected with the theory of functions of a complex variable, while in three dimensions the Laplace equation is of fundamental importance in the theory of Newtonian gravitational potentials. Many significant properties of self-adjoint elliptic equations first discovered in connexion with these and other applications are also valid in an arbitrary number of dimensions, and in curved Riemann spaces. Consequently we shall, using the background of the preceding chapter, develop a 'potential theory' for the equation

$$\Delta u(P) = q(P)u(P),$$

which is the most general form for a self-adjoint elliptic equation.

For this equation there is defined a Dirichlet integral (4.3.24) from which may be derived basic properties of the differential equation when the coefficient $q(P)$ is positive. We shall discover that the sign of this coefficient has a considerable influence on the properties of solutions; so that the Laplace equation, for which $q(P) \equiv 0$, is somewhat exceptional A direct treatment of harmonic functions is therefore postponed to Chapter VIII.

The theory of potential derives its name from the physical interpretation of certain types of solution functions which are built up from solutions having singularities, such as (4.4.1). In the third and following sections we study these potential functions, which are solutions of the above equation or of the associated non-homogeneous equation of Poisson. This leads us in turn to the formulation of boundary value problems appropriate to such equations.

The existence theorems of Chapter I will apply to our present equation only if all coefficients a^{ik} and q are analytic; even then, we are assured of local solutions only. Later it will be shown that solutions in the large exist, and under hypotheses much less restrictive on the coefficient functions. It seems to be characteristic of elliptic equations, that the solutions have much the same degree of regularity as do the coefficients.

5.1. The Dirichlet integral. From § 4.3 we observe that for our self-adjoint equation the appropriate form of Green's first formula (4.3.22) is

$$\int_D v[\Delta u - qu]\, dV + \int_D [\nabla u \cdot \nabla v + quv]\, dV = \int_B v \frac{\partial u}{\partial n}\, dS. \qquad (5.1.1)$$

Now the Dirichlet integral

$$E(u, v) = \int_D [\nabla u \cdot \nabla v + quv]\, dV \qquad (5.1.2)$$

is positive definite if $q > 0$ in V, in the sense that $E(u, u) \geqslant 0$, with $E(u, u) = 0$ only if $u \equiv 0$. If $q = 0$, then $E(u, u) = 0$ implies that $\nabla u = 0$, i.e. that $u = $ constant in D.

We shall therefore assume that q is non-negative in the region under consideration, viz.

$$q(P) \geqslant 0. \qquad (5.1.3)$$

Green's formula (5.1.1) now enables us to deduce striking limitations on the behaviour of solutions of $\Delta u = qu$ in D. Let us set $v = u$ in (5.1.1) and suppose that the behaviour of u on B is such that the boundary integral vanishes. Then, if u is a solution of the differential equation, we must have

$$E(u, u) = 0 \qquad (5.1.4)$$

so that u must be identically zero; unless $q \equiv 0$, in which case u may be a non-zero constant. Now the boundary integral

$$\int_B u \frac{\partial u}{\partial n}\, dS \qquad (5.1.5)$$

vanishes, for instance, if either

(a) $u = 0$ on B, and $\partial u/\partial n$ exists and is bounded in a neighbourhood of B,
(b) $\partial u/\partial n = 0$ on B, or if
(c) $u = 0$ on a portion of B while $\partial u/\partial n$ is zero on the remaining portion.

If $q \geqslant 0$ and $q \not\equiv 0$, any solution of $\Delta u = qu$ satisfying one of (a), (b), or (c) is identically zero in D. If $q \equiv 0$, the same remark holds, except that in case of condition (b), u may be a non-zero constant.

These results can be interpreted as statements that solutions of the differential equation having certain assigned properties on the boundary of a region are uniquely determined. For example, if u is a solution of $\Delta u = qu$, $q \geqslant 0$, and u has assigned values $f(p)$ on B:

$$u(p) = f(p), \qquad (5.1.6)$$

then u is uniquely determined. For suppose that u_1 is any other function with these same properties, then $u - u_1$ is a solution of the linear equation

$\Delta u = qu$, and is zero on B. Hence $u - u_1 \equiv 0$, so that u_1 is necessarily identical with u. The problem of showing that there exists a solution u with assigned values on B is known as Dirichlet's problem. Thus if $q \geqslant 0$, a solution of Dirichlet's problem is unique. If we can find a solution for Dirichlet's problem, the problem will be correctly set, since it can have at most one solution.

Similarly, the problem of finding a solution u of the differential equation which satisfies on the boundary a condition of the form

$$\frac{\partial u(p)}{\partial n} = g(p) \tag{5.1.7}$$

is important in potential theory, and is known as Neumann's problem. Our theorem shows easily that, if $q \geqslant 0$, $q \not\equiv 0$, a solution of the Neumann problem is unique. However, if $q \equiv 0$, the reasoning shows only that the difference of two solutions is a constant. Since any constant is a solution of the Neumann problem with $g(p) = 0$, it is clear that in this case the solution of the Neumann problem is determined up to an additive constant. This lack of determination of the solution is an exceptional phenomenon; it is our simplest example of an eigenfunction, a solution of the differential equation which also satisfies the homogeneous boundary condition.

Moreover, in this case $q \equiv 0$, the Neumann problem cannot be solved for arbitrary (even continuous or smooth) functions $g(p)$. For, if $q \equiv 0$, we have, on setting $v = 1$ in (5.1.1),

$$\int_B \frac{\partial u}{\partial n} \, dS = E(1, u) + \int_D 1 \, \Delta u \, dV = 0. \tag{5.1.8}$$

Equation (5.1.8) states that the assigned values of the normal derivative of a harmonic function must have a zero mean value over the surface. This is the Gauss integral theorem of potential theory, and it shows that if the Neumann problem is to be solvable, we must have

$$\int_B g(p) \, dS = 0. \tag{5.1.9}$$

Thus these two phenomena occur together when $q(p) \equiv 0$, namely, (a) the existence of an eigenfunction, and (b) the necessity of a condition on the data of the problem. Later we shall see, by means of integral equations, how these two matters are bound up together.

There is a third type of condition, known as a mixed boundary condition, arising from physical problems such as that of heat conduction. The boundary value problem associated with this condition is known as Robin's problem. The condition itself is

$$\frac{\partial u(p)}{\partial n} + h(p)u(p) = f(p) \quad (h(p) \geqslant 0, h(p) \not\equiv 0). \tag{5.1.10}$$

Thus $h(p)$ is a continuous non-negative function which does not vanish identically (or we would be back to (5.1.7)). We can prove that the solution of Robin's problem, when it exists, is unique if $q(p) \geqslant 0$. For the difference u between two solutions satisfies the homogeneous boundary condition

$$\frac{\partial u(p)}{\partial n} + h(p)u(p) = 0. \tag{5.1.11}$$

Then, from (5.1.1) we have, since $\Delta u = qu$,

$$E(u, u) = \int_B u\frac{\partial u}{\partial n} \, dS = - \int_B hu^2 \, dS \leqslant 0. \tag{5.1.12}$$

Since $E(u, u)$ is non-negative, the only possibility is now that $E(u, u) = 0$, and then the integral on the right of (5.1.12) is zero also. Hence u is zero at those points of B where $h \neq 0$, and finally, therefore, u must be identically zero by the positivity of the Dirichlet integral. This reasoning holds only if $q(p)$ and $h(p)$ are non-negative.

We include in this section the following reciprocal theorem for two solutions u_1 and u_2 of the differential equation $\Delta u = qu$ in D. If u_1 and u_2 satisfy $\Delta u = qu$ in D and are continuous on B, then

$$\int_B u_1\frac{\partial u_2}{\partial n} \, dS = \int_B u_2\frac{\partial u_1}{\partial n} \, dS. \tag{5.1.13}$$

This formula is an immediate consequence of Green's formula (5.1.1) for the differential equation in question, and of the fact that

$$E(u_1, u_2) = E(u_2, u_1).$$

Note that this theorem holds for unrestricted values of the function $q(P)$.

Returning now to our hypothesis that the function $q(P)$ is positive, we recall that the Dirichlet integral is then positive definite. In many physical applications in which the argument function u of $E(u)$ represents a potential function of some kind or a temperature distribution, the Dirichlet integral itself may be interpreted as the energy associated with the configuration of the system. Many of the physical principles which determine the configuration of physical systems can be stated in this way: the energy shall be a minimum. Now it is also possible to describe the configuration by means of a differential equation. That these two ways of determining the state of a physical system lead to the same result may be seen by the 'minimum principle': *Of all functions $u(P)$ of class C^1 defined on D and having given boundary values $f(p)$ on B, the solution u_0 of $\Delta u = qu$ with these boundary values has (if it exists) the smallest Dirichlet integral.*

This can be proved in the following way: let $u_0(P)$ be the solution of the differential equation with boundary values $f(p)$, and let $u(P)$ be any other (C^1) function with these boundary values. Then the difference

$$v(P) = u(P) - u_0(P)$$

has boundary value zero, and is of class C^1. Furthermore, we have $u(P) = u_0(P) + v(P)$, so that

$$E(u, u) = E(u_0, u_0) + 2E(u_0, v) + E(v, v).$$

From Green's formula we see that

$$E(u_0, v) = \int_B v \frac{\partial u_0}{\partial n} \, dS - \int_D v[\Delta u_0 - q u_0] \, dV.$$

The surface integral is zero since $v = 0$ on B, and the volume integral is zero because u_0 is a solution of the differential equation in D. Therefore $E(u_0, v) = 0$, and

$$E(u, u) = E(u_0, u_0) + E(v, v) > E(u_0, u_0)$$

since $E(v, v) > 0$, v being different from zero. This establishes the minimal property of solutions of the differential equation in relation to the Dirichlet integral.

Exercise 1. Prove that the minimum property holds if $q(P)$ is non-negative or zero.

Exercise 2. If $u(P)$ is a function such that $E(u, u)$ is minimal among those functions with boundary values equal to $u(p) = f(p)$ show that, if $v(p) = 0$ on B, then

(1) $E(u, v) = 0$,

(2) $\int_D vL[u] \, dV = 0.$

5.2. A maximum principle. In the previous section we saw that if $q(P)$ is non-negative, it is possible to deduce from the Dirichlet integral quite sharp limitations on the behaviour of solutions. Now if $q(P)$ is actually positive, $q(P) > 0$, certain other limitations can be found directly from the differential equation. In fact, we have the 'maximum principle': *If $(qP) > 0$, a solution of $\Delta u = qu$ can have no positive maximum or negative* *minimum in the interior of the region in which the differential equation is satisfied.*

The proof depends on the fact that at a maximum point the second derivatives $\dfrac{\partial^2 u}{(\partial x^i)^2}$ are non-positive, while the first derivatives are all zero. Let us suppose that $u(P)$ has a maximum at P, and let us choose a local Cartesian coordinate system at P such that all the non-diagonal components of a vanish (32, p. 59). Then, for the invariant scalar Δu, we have, in this coordinate system,

$$\Delta u = \sum_i a^{ii} \frac{\partial^2 u}{(\partial x^i)^2} \leqslant 0, \qquad (5.2.1)$$

since the a^{ii} are all positive and the second derivatives non-positive. If, then, the differential equation $\Delta u = qu$ holds, the product qu must be non-positive. Since q is positive, we see that u itself must be non-positive. This proves the first statement. The absence of negative minima can be shown by changing the sign of u and then applying the above result.

Exercise 1. If $q(P) < 0$ in the domain V, show that solutions of $\Delta u = qu$ have no positive minimum or negative maximum.

Exercise 2. How can Theorem 1 be modified for the inhomogeneous equation
$$\Delta u - qu = f,$$
where f is continuous and bounded in V; thus $|f(P)| < K$?

Exercise 3. Show that
$$u = \sin x^1 \sin x^2 \ldots \sin x^N$$
is a solution of
$$\Delta u = -Nu,$$
in a flat space with Cartesian coordinates x^i. Show that this solution provides a counterexample to the hypothesis that the maximum principle holds for $q < 0$.

A first consequence of this maximum principle is that any solution assumes whatever positive maximum or negative minimum values it may have on the boundary of its domain. From this remark it follows that a solution which vanishes on the boundary is certainly zero throughout. For if such a solution assumed, say, positive values, it would have a positive maximum which is impossible. Thus we have a second proof that the Dirichlet problem for $\Delta u = qu$, $q > 0$, has at most one solution. Another consequence of the theorem is that a solution which is non-negative on the boundary is non-negative in the whole interior. For, if the solution takes on negative values only in the interior, it must have a negative minimum in the interior, and this cannot happen.

Exercise 1. Let $u_1(P)$ be a solution with $u_1(p) = 1$ on B. If $u(P)$ is any solution with $m \leqslant u(p) \leqslant M$ on B, then $mu_1 \leqslant u \leqslant Mu_1$ in D.

Exercise 2. The maximum principle holds for linear second order equations which are not self-adjoint, viz.,

$$\Delta u + \mathbf{b} \cdot \nabla u + cu = 0 \quad (c < 0).$$

Consequently a solution of the Dirichlet problem is unique here also.

Exercise 3. Show that if a maximum principle holds, then solutions of the Dirichlet problem are stable in the sense of § 1.2. Is this also true for Neumann's problem?

5.3. The local fundamental solution. In the preceding chapter we have seen that an elliptic equation with constant coefficients has a fundamental solution or fundamental singularity, such as the solution r^{-N+2} of Laplace's equation. Equations with variable coefficients also have solutions with quite similar properties. Thus we shall use the term fundamental solution to denote a function of two points $\gamma(P, Q)$ which satisfies the following conditions:

(1) As a function of the argument point P, $\gamma(P, Q)$ is a regular solution of the differential equation $Lu = 0$, except when $P = Q$.

(2) At the parameter point Q, $\gamma(P, Q)$, regarded as a function of P, has the singularity

$$\gamma(P, Q) \sim \frac{1}{\omega_N(N-2)} s(P, Q)^{-N+2}, \tag{5.3.1}$$

where $N > 2$, and $\qquad \omega_N = \dfrac{2\pi^{\frac{1}{2}N}}{\Gamma(\frac{1}{2}N)}$

is the solid angle at a point of N-dimensional space.

The reason for the particular choice of the numerical factor in (5.3.1) will shortly appear. If any regular solution $u(P)$ of the differential equation is added to $\gamma(P, Q)$, the sum is again a fundamental solution.

The fundamental solution has a simple physical interpretation. Imagine a Newtonian field of gravity, or an electrostatic field, specified by a potential function $u(P)$ which satisfies the differential equation $Lu = 0$ under consideration. Let a unit mass or charge be concentrated at a given point Q. Then the potential due to this source is $-\gamma(P, Q)$, where $\gamma(P, Q)$ is a

fundamental solution. When the mass or charge is distributed continuously with volume density $f(P)$, the potential function satisfies the Poisson equation
$$Lu = \Delta u - qu = f(P). \tag{5.3.2}$$

Expressing the solution of this equation as the sum of contributions of the mass or charge distribution, we are led to consider an integral of the form

$$u(P) = -\int \gamma(P, Q) f(Q) \, dV_Q. \tag{5.3.3}$$

Let us now investigate this relationship more closely. We will show that, provided the domain is taken sufficiently small, the Poisson equation (5.3.2) has a solution of the form (5.3.3), and that the function $\gamma(P, Q)$ appearing there is a fundamental solution.

Though we have not yet shown that a fundamental solution exists, we can at least claim to know a function which for P and Q close together is a good approximation to one, namely, the 'parametrix'

$$\omega(P, Q) = \frac{s(P, Q)^{-N+2}}{\omega_N(N-2)} = \frac{\Gamma^{-\frac{1}{2}N+1}}{\omega_N(N-2)}. \tag{5.3.4}$$

The geodesic distance is defined if P and Q are sufficiently close, which we shall assume. From (4.5.23) we see that

$$\Delta\omega(P, Q) = \frac{\Delta(\Gamma^{-\frac{1}{2}N+1})}{\omega_N(N-2)} = \frac{\Gamma^{-\frac{1}{2}N}}{2\omega_N}\{-2N+\Delta\Gamma\}$$

$$= \frac{\Gamma^{-\frac{1}{2}N}}{2\omega_N}2s\frac{\partial\log\sqrt{a}}{\partial s} = O(\Gamma^{-\frac{1}{2}N+1}),$$

in view of (4.5.31). Therefore, if we define $q(P, Q)$ as the amount by which $\omega(P, Q)$ fails to be an exact solution:

$$q(P, Q) = L_P\omega(P, Q) = \Delta_P\omega(P, Q) - q(P)\omega(P, Q), \tag{5.3.5}$$

we see that the singularity of $q(P, Q)$ as $P \to Q$ is also of order $N-2$, and so $q(P, Q)$ is integrable as a function of either argument separately.

Based on the approximate fundamental solution $\omega(P, Q)$ we have the approximate solution of the Poisson equation

$$U(P) = \Omega f(P) = -\int \omega(P, Q) f(Q) \, dV_Q. \tag{5.3.6}$$

This equation defines an integral operator Ω; for the present we leave the domain D of integration unspecified. To see how good an approximation is furnished by (5.3.6) we should apply to it the differential operator L.

Because of the singularity of $\omega(P, Q)$ as $Q \to P$, we cannot perform the differentiation formally under the integral sign. Let us write (5.3.6) as the sum of an integral U_1 over $K_\epsilon(P)$ and an integral U_2 over $D - K_\epsilon(P)$, where $K_\epsilon(P)$ is a geodesic sphere of radius ϵ about P. Then $\omega(P, Q)$ is bounded in $D - K_\epsilon(P)$, and we may differentiate U_2 formally under the integral sign. We now calculate the Laplacian of the integral U_1, at least in the limit as $\epsilon \to 0$.

In a coordinate system valid in a neighbourhood of P, let the coordinates of P be x^i and let those of Q be ξ^i. The quantity $\Gamma(P, Q)$, regarded as a function of Q, has as Taylor expansion about P to the third order,

$$\Gamma(P, Q) = a_{ik}(P)(x^i - \xi^i)(x^k - \xi^k) + A_{ijk}(x^i - \xi^i)(x^j - \xi^j)(x^k - \xi^k).$$

Here the $a_{ik}(P)$ are metric tensor components and the A_{ijk} are bounded and differentiable. Thus we find

$$\frac{\partial \Gamma}{\partial x^i} = 2a_{ik}(P)(x^k - \xi^k) + B_{ijk}(x^j - \xi^j)(x^k - \xi^k),$$

and
$$\frac{\partial \Gamma}{\partial \xi^i} = -2a_{ik}(P)(x^k - \xi^k) + C_{ijk}(x^j - \xi^j)(x^k - \xi^k),$$

where the B_{ijk} and C_{ijk} are also bounded. Since the coordinate intervals $x^i - \xi^i$ have order of magnitude s, we see that

$$\frac{\partial \Gamma}{\partial \xi^i} = \frac{\partial \Gamma}{\partial x^i} + O(s^2).$$

Thus the error term is of order s^2. Now also

$$\frac{\partial \omega(P, Q)}{\partial \xi^i} = -\frac{\Gamma^{-\frac{1}{2}N}}{\omega_N} \frac{\partial \Gamma}{\partial \xi^i}$$

$$= +\frac{\Gamma^{-\frac{1}{2}N}}{\omega_N} \frac{\partial \Gamma}{\partial x^i} + O(s^{-N+2}) = -\frac{\partial \omega(P, Q)}{\partial x^i} + O(s^{-N+2}).$$

(5.3.7)

It follows from (5.3.7) that

$$\frac{\partial U_1(P)}{\partial x^i} = -\int_{K_\epsilon(P)} \frac{\partial}{\partial x^i} \omega(P, Q) f(Q) \, dV_Q$$

$$= +\int_{K_\epsilon(P)} \frac{\partial}{\partial \xi^i} \omega(P, Q) f(Q) \, dV_Q + E_1,$$

(5.3.8)

where
$$|E_1| \leqslant K \int_{K_\epsilon(P)} s^{-N+2} |f(Q)| \, dV_Q.$$

Taking geodesic polar coordinates centred at P (§ 4.5, Ex. 4), we find

$$|E_1| \leqslant K \int_{\omega_N} \int_0^\epsilon s^{-N+2} |f(Q)| s^{N-1} \, ds d\Omega$$

$$\leqslant \tfrac{1}{2} K \omega_N \max |f(Q)| \epsilon^2 = O(\epsilon^2).$$

The integral on the right in (5.3.8), after an integration by parts, becomes

$$\int_{K_\epsilon(P)} \frac{\partial}{\partial \xi^i} (\omega(P,Q) f(Q) \sqrt{a}) \, d\xi^1 ... d\xi^N - \int_{K_\epsilon(P)} \omega(P,Q) \frac{\partial}{\partial \xi^i} (f(Q) \sqrt{a}) \, d\xi^1 ... d\xi^N.$$

Here we must assume that $f(Q)$ is once differentiable. The first of these integrals may be integrated out to a surface integral, while the second is seen to be of order ϵ^2, with first derivatives of order ϵ. Thus, from (4.3.4), we find

$$\frac{\partial U_1(P)}{\partial x^i} = + \int_{s=\epsilon} \omega(P,q) f(q) \sqrt{a} \, D_i \, dt^1 ... dt^{N-1} + O(\epsilon^2),$$

and this integral itself is of order ϵ as $\epsilon \to 0$. We have denoted points on the surface $s = \epsilon$ by a small letter variable.

Near the point P we have, since the first derivatives of $U_1(P)$ are of order ϵ,

$$\Delta U_1(P) = a^{ik}(P) \frac{\partial}{\partial x^k} \int_{s=\epsilon} \omega(P,q) f(q) \sqrt{a} \, D_i \, dt^1 ... dt^{N-1} + O(\epsilon)$$

$$= -a^{ik}(P) \int_{s=\epsilon} \frac{\partial \omega(P,q)}{\partial \xi^k} f(q) \sqrt{a} \, D_i \, dt^1 ... dt^{N-1} + O(\epsilon)$$

$$= - \int_{s=\epsilon} a^{ik}(q) \frac{\partial \omega(P,q)}{\partial \xi^k} \frac{D_i}{D} f(Q) \sqrt{a} \, D dt^1 ... dt^{N-1} + O(\epsilon),$$

using (4.3.8) and the continuity of the a^{ik}.

From (4.3.11) we see that

$$\Delta U_1(P) = - \int_{s=\epsilon} \frac{\partial \omega(P,q)}{\partial n} f(q) \, dS_q + O(\epsilon), \qquad (5.3.9)$$

where n denotes the outward normal to the sphere. Thus $\partial s(P, q)/\partial n = 1$, and

$$\Delta U_1(P) = - \int_{s=\epsilon} \frac{1}{\omega_N(N-2)} \frac{\partial s^{-N+2}}{\partial n} f(q) \, dS_q + O(\epsilon)$$

$$= \frac{1}{\omega_N} \int_{s=\epsilon} s^{-N+1} f(q) s^{N-1} \, d\Omega + O(\epsilon) \qquad (5.3.10)$$

$$= \frac{1}{\omega_N} \int_{s=\epsilon} f(q) \, d\Omega + O(\epsilon) \rightarrow f(P),$$

as $\epsilon \rightarrow 0$, since $f(P)$ is continuous and $\int d\Omega = \omega_N$.

Finally, we see that in the limit as $\epsilon \rightarrow 0$,

$$L\Omega f(P) = LU_1 + LU_2$$

$$= f(P) - \int L_P \omega(P, Q) f(Q) \, dV_Q \qquad (5.3.11)$$

$$= f(P) - \int q(P, Q) f(Q) \, dV_Q.$$

This equation holds for all once differentiable functions $f(P)$. Let us define a second integral operator, based upon the kernel $q(P, Q)$:

$$\Phi f(P) = \int q(P, Q) f(Q) \, dV_Q. \qquad (5.3.12)$$

Then we may write (5.3.11) in the form

$$L\Omega f(P) = f(P) - \Phi f(P). \qquad (5.3.11)$$

We are now ready to find a solution of the Poisson equation $Lu = f(P)$. The term $\Phi f(P)$ in (5.3.11) is a measure of the error incurred by the use of the parametrix. Let us attempt to compensate this by taking as trial solution of the Poisson equation

$$u(P) = \Omega v(P),$$

where $v(P)$ is a new function which is to be determined. We have, then,

$$Lu(P) = L\Omega v(P) = v(P) - \Phi v(P),$$

which quantity we shall equate to the density $f(P)$. Thus we find for $v(P)$ the integral equation

$$v(P) - \Phi v(P) \equiv v(P) - \int q(P, Q) v(Q) \, dV_Q = f(P), \qquad (5.3.13)$$

wherein the unknown function $v(P)$ appears beneath the sign of integration. Symbolically, (5.3.13) is $(1 - \Phi)v = f$; and it has the symbolic solution

$$v(P) = (1 - \Phi)^{-1} f(P) = \sum_{p=0}^{\infty} \Phi^p f(P), \qquad (5.3.14)$$

where Φ^p denotes the integral operator consisting of p applications of Φ.

We shall now demonstrate that if the domain of integration D is chosen sufficiently small, the series (5.3.14) converges and provides a true solution of the integral equation.

The singularity of $q(P, Q)$ is integrable; that is,

$$\int |q(P, Q)|\, dV_Q,$$

taken over a sphere $K_\epsilon(P)$ of radius ϵ about P, exists, and tends to zero with ϵ. Let us choose as domain of integration a region D of diameter so small that the above integral is less than a number K less than unity for all points P in D. Then, if $|f(P)|$ has a bound M in D,

$$|\Phi f(P)| = \left| \int q(P, Q) f(Q)\, dV_Q \right|$$

$$\leqslant M \int |q(P, Q)|\, dV_Q \leqslant MK,$$

and, by iteration,

$$|\Phi^p f(P)| \leqslant MK^p \qquad (p = 1, 2, 3, ...).$$

That is, the terms of the series (5.3.14) are bounded by the terms MK^p of a geometrical series with common ratio $K < 1$. Hence the series (5.3.14) converges uniformly for P in the region D. It is now easy to verify that $v(P)$ is a solution function for the integral equation.

Returning to our original problem, we find that

$$u(P) = \Omega v(P) = \Omega \sum_{p=0}^{\infty} \Phi^p f(P) \qquad (5.3.15)$$

is a solution of the Poisson equation in the small domain D. The reader will easily verify that the rather complicated operation performed on $f(P)$ in (5.3.15) is again an integral operator with domain of integration D. The kernel of this integral operator is given by a series

$$\gamma(P, Q) = \omega(P, Q) + \int \omega(Q, R) q(Q, R)\, dV_R + ..., \qquad (5.3.16)$$

which is asymptotic to the parametrix $\omega(P, Q)$ when $P \to Q$ (cf. § 6.4). This function $\gamma(P, Q)$, as we might anticipate, is a fundamental solution in the small domain D.

In order to verify this fact, we must show that $\gamma(P, Q)$, as function of P, is a solution of the differential equation for $P \neq Q$. Suppose, to the contrary, that

$$L_P \gamma(P, Q) > 0 \qquad (5.3.17)$$

for points P and Q which are not the same. Let $f(R)$ be a non-negative, continuous function which is zero at P and everywhere except in a neighbourhood N of Q. Let this neighbourhood be chosen so small that (5.3.17) holds at each point R of N. Then we have

$$L \int \gamma(P, R) f(R) \, dV_R = 0,$$

since the integral is a solution of the Poisson equation, and since $f(P)$ is zero. But, since $f(R)$ vanishes near P, we can differentiate under the integral sign and thus

$$L \int \gamma(P, R) f(R) \, dV_R = \int_N L\gamma(P, R) f(R) \, dV_R > 0,$$

and this is a contradiction. Hence $L_P \gamma(P, R) = 0$, and we have shown that $\gamma(P, R)$ is a local fundamental solution.

A study of the convergence of the series (5.3.16), which we leave as an exercise, shows that $\gamma(P, Q)$ possesses as many derivatives with respect to P as the parametrix $\omega(P, Q)$, that is, as many as the geodesic distance $s(P, Q)$. The above construction clearly requires that $s(P, Q)$ should be twice differentiable as a function of P or of Q, a circumstance which will be fulfilled if the coefficients a^{ik} of the differential equation are four times differentiable. When the differential equation is analytic, so are the metric, the parametrix, and hence also the local fundamental solution.

No restriction of sign on the coefficient function $q(P)$ is needed in this work, so we may conclude that every self-adjoint second order elliptic equation has a local fundamental solution, defined in a finite but small domain.

Exercise 1. Using the notation of the integral operators, write out the full series for $\gamma(P, Q)$ in (5.3.16). Show from (5.3.11) that $L_P \gamma(P, Q) = 0$, $P \neq Q$.

Exercise 2. Find a suitable parametrix for an equation

$$\Delta u + \mathbf{b} . \nabla u + cu = f,$$

which is not self-adjoint. *Hint*: Take local Cartesian coordinates at Q and follow the reduction in § 4.4 of the equation with constant coefficients to self-adjoint form.

Exercise 3. Show that the equation of Exercise 2 has solutions in sufficiently small domains.

5.4. Volume and surface potentials. The existence of a fundamental solution enables us to construct solutions of the Poisson equation $Lu = f(P)$ in the form of volume potentials. At points of empty space—

that is, where $f(P)$ vanishes—these functions satisfy the homogeneous equation $Lu = 0$. Our purpose in this section is to use Green's formula, in conjunction with a fundamental solution, to study the potentials of distributions in space and on surfaces bounding regions of space. We shall therefore assume that a fundamental solution exists and is defined in the whole of the region in which we are interested.

Suppose that $u(P)$ satisfies the Poisson equation

$$Lu(P) = f(P), \tag{5.4.1}$$

in the interior of a domain D. Denoting the variable point of integration by Q, we apply Green's second formula (4.3.19) to the functions $u(Q)$ and $\gamma(Q, P)$. That is, we take Q as argument point and P as parameter point, so that $L_Q \gamma(Q, P) = 0$ for $Q \neq P$. Since $\gamma(Q, P)$ has an infinity for $Q = P$, we apply the formula to the domain D from which a spherical neighbourhood $K_\epsilon(P)$ has been removed, and then take the limit as $\epsilon \to 0$. In that domain we have $L_Q \gamma(Q, P) = 0$, and also (5.4.1) holds, so that we get

$$\int_{D-K_\epsilon} \gamma(Q, P) f(Q)\, dV_Q = \int_{B-S_\epsilon} \left\{ \gamma(q, P) \frac{\partial u(q)}{\partial n} - u(q) \frac{\partial \gamma(q, P)}{\partial n} \right\} dS_q. \tag{5.4.2}$$

Here B denotes the boundary of D, and S_ϵ the boundary of K_ϵ.

Now we shall make the transition to the limit. In view of (5.3.1) the volume integral converges like

$$\int_0^\epsilon s^{-N+2}\, dV = \int_0^\epsilon s^{-N+2} s^{N-1}\, ds\, d\Omega = O(\epsilon^2)$$

as $\epsilon \to 0$. Now the surface integral consists of two parts; with one only of these are we concerned here, namely, the integrals over the small spherical surface S_ϵ about P. In the spherical coordinates,

$$\int_{S_\epsilon} \gamma(q, P) \frac{\partial u(q)}{\partial n}\, dS_q \sim \frac{+1}{\omega_N(N-2)} \int_{s=\epsilon} s^{-N+2} \frac{\partial u}{\partial n} s^{N-1}\, d\Omega = O(\epsilon) \quad (\epsilon \to 0),$$

so the first term in the surface integral over S_ϵ converges to zero as limit. In the second term appears the quantity

$$\frac{\partial \gamma(q, P)}{\partial n_q} \sim \frac{1}{\omega_N(N-2)} \frac{\partial}{\partial n} s^{-N+2} + \dots$$

$$= \frac{-1}{\omega_N(N-2)} \frac{\partial}{\partial s} s^{-N+2} + \dots.$$

The change of sign takes place because n, the outward normal to $D-K_\epsilon$, is the inward normal to K_ϵ. The normal is also a geodesic line through the centre point P of the sphere $K_\epsilon(P)$.

The second term on the right of (5.4.2) may now be transformed in the following way (cf. (5.3.9)):

$$\int_{S_\epsilon} u(q) \frac{\partial \gamma(q, P)}{\partial n} \, dS_q = -\int_{S_\epsilon} u(q) \frac{\partial}{\partial s} \left(\frac{s^{-N+2}}{\omega_N(N-2)} \right) dS + O(\epsilon)$$

$$= \frac{1}{\omega_N} \int_{s=\epsilon} u(q) s^{-N+1} s^{N-1} \, d\Omega + O(\epsilon) \to u(P). \tag{5.4.3}$$

The limit of the integral over the small sphere about P is precisely the value of the function $u(P)$. When we take account of this and rearrange the terms in (5.4.2), we find the important formula

$$\alpha u(P) = -\int_D \gamma(Q, P) f(Q) \, dV_Q + \int_B \left\{ \gamma(q, P) \frac{\partial u(q)}{\partial n} - u(q) \frac{\partial \gamma(q, P)}{\partial n} \right\} dS_q. \tag{5.4.4}$$

Here α is the number unity when P lies in the interior of D. From the discussion of formula (5.4.2), it is seen that α should be taken equal to zero when P lies outside of the domain D. A further possibility may be considered, namely, that P is a boundary point of D and so lies on B. To derive our formula in this case, we should apply Green's theorem to D after removing from D a hemispherical neighbourhood of P. The limiting process would then follow through as before, except that a factor of $\frac{1}{2}$ should appear in the result of (5.4.3). In (5.4.4), therefore, we have

$$\alpha = \begin{cases} 1 & P \text{ in } D, \\ \frac{1}{2} & P \text{ on } B, \\ 0 & P \text{ outside } D. \end{cases} \tag{5.4.5}$$

The basic formula (5.4.4) is useful in the study of solutions of the homogeneous equation $Lu = 0$, in which case the function $f(P)$ vanishes. Then we have, for P in D,

$$u(P) = \int_B \left\{ \gamma(q, P) \frac{\partial u(q)}{\partial n} - u(q) \frac{\partial \gamma(q, P)}{\partial n} \right\} dS_q. \tag{5.4.6}$$

Thus any solution function can be represented in this way if the values of $u(q)$ and the normal derivative $\partial u(q)/\partial n$ are given on the boundary B. However, the uniqueness theorems of § 5.1 show that values of $u(q)$ alone are sufficient to determine a solution uniquely, at least when $q(P) > 0$. Thus we cannot assign arbitrary values to both $u(q)$ and $\partial u(q)/\partial n$ in (5.4.6). We have observed that when the differential equation is analytic, it possesses an analytic local fundamental solution. In this case we shall apply (5.4.6) for a sufficiently small region D enclosing the point P. The right-hand side is then an analytic function of P in the interior of D, and we conclude that *every solution of an analytic elliptic differential equation* $L[u] = 0$ *is itself analytic.*

We shall make further use of the formula (5.4.4) in connexion with the various terms appearing on the right-hand side. In all of the applications the fundamental solution which appears in this formula will be defined in the large, and will satisfy the homogeneous differential equation $L\gamma = 0$ as a function of each of the variables P and Q. That is, the parameter point will also be an argument point. Clearly the symmetric solutions (4.4.1), (4.4.2), and (4.4.3) have this property. We shall therefore assume that the fundamental solution in (5.4.4) satisfies $L_P\gamma(P, Q) = 0$ and also $L_Q\gamma(P, Q) = 0$, $P \neq Q$. Thus we may now write either $\gamma(P, Q)$ or $\gamma(Q, P)$ in that formula.

The three integrals on the right in (5.4.4) are the potentials which we now examine. First is the volume potential

$$U(P) = -\int_D \gamma(P, Q)f(Q)\, dV_Q. \tag{5.4.7}$$

This integral converges uniformly with respect to P and so defines a continuous function. Indeed, the calculations leading to (5.3.11) show that if $f(P)$ is once differentiable, then $U(P)$ possesses second derivatives and that $LU(P) = f(P)$. The single layer surface potential

$$V(P) = \int_B \gamma(P, q)g(q)\, dS_q, \tag{5.4.8}$$

of single layer density $g(q)$ on the surface B, appears in (5.4.4) with density the value of the normal derivative of $u(P)$. Finally, the double layer surface potential

$$W(P) = -\int_B \frac{\partial\gamma(P, q)}{\partial n} h(q)\, dS_q, \tag{5.4.9}$$

with double layer density $h(q)$, is present in (5.4.4) with density equal to the values of $u(P)$ on the boundary surface. The two surface potentials $V(P)$ and $W(P)$ are both solutions of the homogeneous differential equation $Lu(P) = 0$ for P in D. For if P does not lie on B, the argument points P and q of $\gamma(P, q)$ never coincide, so that $L_P\gamma(P, q) = 0$ for all q on B.

We note that $u(P)$ in (5.4.4) may be any twice-differentiable function; then $f(P)$ is regarded as defined in (5.4.1) by applying the operator L to $u(P)$. Thus any suitably differentiable function defined in a domain D may, by the formula (5.4.4), be expressed as a sum of a volume potential, a single layer, and a double layer potential.

The single and double layer potentials are solutions of the equation at all points not on the boundary surface B upon which the layer densities are imagined to be situated. For our later work it is necessary to establish certain discontinuity properties of these potentials, and of their normal derivatives, as the argument point crosses the surface B. Let us denote by (u) the difference of values of a function u discontinuous across B, the argument point being understood to cross B from the interior to the exterior of D. We shall also use a $+$ subscript to denote limits from the exterior, $-$ for limits from the interior, and a small letter to indicate an argument point situated upon the boundary.

From formula (5.4.4) we see that the left-hand side has a discontinuity of $-u(p)$ across the boundary, p being the point of crossing. Now, as regards the terms on the right, the volume potential, as well as its first partial derivatives, is continuous across B, since the integral itself is absolutely and uniformly convergent (together with its first derivatives in which the integrand is $O(s^{-N+1})$), for P in a suitable neighbourhood of B. Similarly, we see that the single layer is defined by an integral whose integrand is $O(s^{-N+2})$, and is therefore absolutely and uniformly convergent for Q in a neighbourhood of B. Since the limit of a uniformly convergent sequence of functions is continuous, the single layer $V(P)$ itself is continuous across B. Of the three terms on the right in (5.4.4), therefore, only the double layer can be discontinuous. In fact, the discontinuity of the double layer must be equal to the negative of the discontinuity of the left-hand side, namely, $-h(p)$:

$$W_+(p)-W_-(p) = \left(- \int_B h(q)\frac{\partial\gamma(P,q)}{\partial n_q} dS_q\right) = -h(p). \qquad (5.4.10)$$

Furthermore, we see from (5.4.5) that the value of this potential on the boundary is the mean of its limiting values from the two sides, as the same is true for the left-hand side of (5.4.4). Therefore (Fig. 5),

$$W_+(p)+\tfrac{1}{2}h(p) = W(p) = W_-(p)-\tfrac{1}{2}h(p). \qquad (5.4.11)$$

This formula will later be applied in connexion with the formulation of boundary value problems by means of integral equations.

Also necessary for this purpose is a similar knowledge of the behaviour of the normal derivatives of the single and double layers. Choosing first the single layer, let us note that we can derive the behaviour across the boundary of its normal derivative by means of the behaviour of the double

FIG. 5. Behaviour of surface layers near the boundary.

layer itself. It is clear that the actual discontinuity of the double layer arises from the leading term of the fundamental singularity, which has the form

$$\gamma(P, Q) = \frac{s^{-N+2}}{\omega_N(N-2)}[1+sA(P, Q)],$$

where $A(P, Q)$ is continuously differentiable for small s. Now

$$\frac{\partial \gamma(P, Q)}{\partial n_P} = -\frac{s^{-N+1}}{\omega_N}\frac{\partial s}{\partial n_P}+O(s^{-N+2})$$

$$= +\frac{s^{-N+1}}{\omega_N}\frac{\partial s}{\partial n_Q}+O(s^{-N+2}) = -\frac{\partial \gamma(P, Q)}{\partial n_Q}+O(s^{-N+2}), \quad (5.4.12)$$

if P and Q lie on the normal at a point of a given surface. Therefore we have, formally,

$$\left(\frac{\partial V(P)}{\partial n_P}\right) = \left(\frac{\partial}{\partial n_P}\int g(q)\gamma(P, q)\, dS_q\right)$$

$$= \left(\int g(q)\frac{\partial \gamma(P, q)}{\partial n_P}\, dS_q\right)$$

$$= \left(-\int g(q)\frac{\partial \gamma(P, q)}{\partial n_q}\, dS_q\right).$$

That is, the normal—or, rather, directional—derivative along the normal at points sufficiently close to B has a discontinuity equal to the jump of a double layer of density $g(p)$. It therefore follows from (5.4.11) that

$$\frac{\partial V(p)}{\partial n_+}+\tfrac{1}{2}g(p) = \frac{\partial V(p)}{\partial n} = \frac{\partial V(p)}{\partial n_-}-\tfrac{1}{2}g(p). \quad (5.4.13)$$

Lastly we must investigate the behaviour of the normal derivative of the double layer, namely, $\partial W/\partial n$, across B. In fact, from (5.4.4),

$$\alpha \frac{\partial u(P)}{\partial n} = -\frac{\partial}{\partial n} \int_D \gamma f\, dV + \frac{\partial W(P)}{\partial n} + \frac{\partial V(P)}{\partial n},$$

where α is defined by (5.4.5). Taking the discontinuity across B of this formula, we find in view of (5.4.13) that

$$-\frac{\partial u(p)}{\partial n} = 0 + \left(\frac{\partial W(p)}{\partial n}\right) + \left\{-\frac{\partial u(p)}{\partial n}\right\}.$$

It follows that $$\frac{\partial W(p)}{\partial n_+} = \frac{\partial W(p)}{\partial n_-};$$ (5.4.14)

the normal derivative of the double layer is continuous across B.

This has been a formal, and therefore tentative, analysis of the discontinuities. A direct study of the integrals, such as is given in (12, vol. iii; 20 or 30), confirms the above conclusions under certain restrictions on the regularity of the layer densities.

5.5. Closed Riemannian spaces. Let D be a bounded domain of the set of coordinate variables x^i $(i = 1,..., N)$, and let us consider a self-adjoint equation of the form

$$\Delta u = q(P)u \qquad (5.5.1)$$

on this domain. As we have seen, the differential equation itself defines a metric tensor a^{ik} from which a Riemannian geometry may be built up on the domain D. We wish to investigate the boundary value problems appropriate to this elliptic differential equation, such as the Dirichlet problem, in which we are to construct a solution of the equation with assigned values on the boundary B of the domain D. Actually, there are two different Dirichlet problems in which boundary values are given on B: the interior problem in which the solution function is defined and satisfies (5.5.1) in D, and the exterior problem in which the solution function is defined everywhere except in D. Between the boundary value problems of these two kinds there subsists a close relationship. For the interior problem, we are given the differential equation only in the domain D itself, without any hypothesis regarding the differential equation or even the structure of the space, elsewhere. We must, therefore, make some definite assumption about the topological character of the complementary part of the space, and also about the differential equation in that region.

In Euclidean geometry, and also in most physical problems, we are accustomed to think of space as infinite and unbounded in extent. There is, however, no reason to consider that our domain D should be regarded as a part of an infinite, open space if an advantage can be gained by supposing otherwise. We will show that any bounded domain D can be regarded as a subdomain of a closed space of finite volume, and that the Riemannian geometry imposed on D by the differential equation may be extended to the whole of this closed space. For example, any region on the surface of the earth is a sub-region of the earth's total surface, which is of finite area and has no boundary.

Let D be the given domain, and B its boundary, and let D_1 be a replica of D which is oppositely oriented, and has boundary B_1 which is a replica of B. That is, to each point P of D there corresponds a point P_1 of D_1, and to each coordinate neighbourhood of D there corresponds a similar coordinate neighbourhood of D_1. In particular, points p of B correspond to points p_1 of B_1. Let us identify corresponding points of B and of B_1, in the sense of regarding them as one and the same. Thus all functions or other quantities which we consider will have one and the same value at p and at p_1. Since p has a half-neighbourhood in D and p_1 an oppositely oriented half-neighbourhood in D_1 we may regard these together as constituting a full neighbourhood of the resulting point P in the combined domains. By this identification D and D_1 are 'sewn' together along their boundaries and form a space F which has no boundary whatever and so is closed. In this process we have made no reference to the metric—indeed the space must be re-garded as flexible and deformable if this identification is to take place in a geometrical sense. The closed space F may be called the double of the domain D. Thus, for example, the double of a circular disk is a two-dimensional sphere, and the double of a ring or annulus is the surface of an anchor-ring, or torus.

The idea of constructing a double, in order to represent a given domain as a subdomain of a closed space, was first used in connexion with Riemann surfaces—that is, for Laplace's equation

$$\frac{\partial^2 u}{\partial x^2} + \frac{\partial^2 u}{\partial y^2} = 0$$

which is so closely related to the theory of analytic functions of a complex variable. Consider in particular the theory of elliptic functions, which was developed in great detail during the nineteenth century. These doubly periodic functions of a complex variable possess a fundamental domain in the shape of a parallelogram whose pairs of opposite sides are respectively equal to the two periods. Since the functions assume equal values at corresponding points of opposite sides, we may identify these points in the manner described above. Thus opposite sides are sewn together; the resulting closed space is a torus. The real or imaginary parts of a doubly periodic analytic function are therefore harmonic functions on a torus.

On the given domain D which is a subdomain of the double F so formed, there is defined a metric tensor a^{ik} and a coefficient function $q(P)$. We may extend the definition of these quantities to the complementary region $\bar{D}$ of F, so that the differential equation (5.5.1) is defined throughout F. This can be done so that the components a^{ik} in F are differentiable an infinite number of times in a suitable set of coordinate variables, provided the same was true originally in D. We see, therefore, that any finite domain D with a Riemannian metric can be represented as a subdomain of a closed Riemannian space F.

Examples of the infinite, open Riemannian space and of the finite closed space are furnished by the spaces of constant curvature, which, next to a flat space, are of the simplest possible type. We know that a Euclidean space E_N, which has zero curvature, has a line element which in spherical polar coordinates may be written

$$ds^2 = dr^2 + r^2\, d\Omega^2, \qquad (5.5.2)$$

where $d\Omega^2$ is the line element on the unit sphere $r = 1$. Thus for three dimensions, in the usual notation,

$$d\Omega^2 = d\theta^2 + \sin^2\theta\, d\varphi^2. \qquad (5.5.3)$$

Also, the surface area of the sphere of radius r in E_N is $\omega_N r^{N-1}$.

A space of constant negative curvature $-R$ has line element

$$ds^2 = dr^2 + R^2 \sinh^2(r/R)\, d\Omega^2, \qquad (5.5.4)$$

where $d\Omega^2$ has the same significance as in (5.5.2). The sphere of radius r now has total surface area $R^{N-1}\sinh^{N-1}(r/R)\omega_N$, which quantity grows exponentially large with r. This infinite space is homogeneous and isotropic in the sense that every direction at every point has the same metrical properties.

The other type of space, with constant positive curvature R, is also homogeneous and isotropic, and has line element

$$ds^2 = dr^2 + R^2 \sin^2(r/R)\, d\Omega^2, \qquad (5.5.5)$$

where $d\Omega^2$ is again the same as in (5.5.2). The sphere of radius r in this space has surface area $R^{N-1}\sin^{N-1}(r/R)\omega_N$; this area increases with r until

$r = \frac{1}{2}\pi R$, and then it decreases to zero at $r = \pi R$. Thus there is only one point in the space at a distance πR from the origin. This point may be interpreted as an antipodal point, by analogy with the antipodal point on a sphere. The unit sphere in E_3, with metric (5.5.3), is a space of constant curvature with radius unity. In fact, any space of constant positive curvature may be regarded as a sphere in a Euclidean (or flat) space of dimension one greater. A space of constant positive curvature is thus closed and has finite volume.

On a closed Riemannian space, Green's formula takes an especially simple form, because there is no boundary and therefore no surface integral. In fact, from (4.3.23) we see that, in this case,

$$\int_F v(\Delta u - qu)\, dV + \int_F \{\nabla u \cdot \nabla v + quv\}\, dV = 0, \tag{5.5.6}$$

for any two suitable functions u and v. In particular, if u is a solution of (5.5.1), we have

$$E(u, u) \equiv \int_F \{(\nabla u)^2 + qu^2\}\, dV = 0, \tag{5.5.7}$$

since the first term of (5.5.6) vanishes. From (5.5.7) we can easily deduce a certain very strong limitation on solutions of $Lu = 0$ in the case when the coefficient $q(P)$ is non-negative, for then the Dirichlet integral (5.5.7) is positive definite. In fact, *if $q(P) \geqslant 0$, $q(P) \not\equiv 0$, the only everywhere regular solution in F of $\Delta u = q(P)u$ is the zero solution.*

When $q \equiv 0$, that is, in the case of Laplace's equation $\Delta u = 0$, a slightly weaker result holds. If u is harmonic and regular in F, the Dirichlet integral $E(u, u)$ is zero, but since $q \equiv 0$ we can only conclude now that $\nabla u = 0$, or that u is a constant. *Thus any harmonic function regular in F is a constant.*

Exercise 1. In a space of constant negative curvature, the function $\gamma(r/R)$, where

$$\gamma(x) = \int_a^x \frac{dt}{\sinh^{N-1}(t)} \qquad \left(x = \frac{r(P, Q)}{R}\right),$$

is a fundamental singularity for the Laplace equation, and is defined throughout the space.

Exercise 2. In a space of constant positive curvature,

$$\gamma(x) = \int_a^x \frac{dt}{\sin^{N-1}(t)} \qquad \left(x = \frac{r(P, Q)}{R}\right)$$

leads to a fundamental singularity for the Laplace equation. Show that this solution has a singularity also when P tends to the point antipodal to Q, and is therefore not strictly a fundamental singularity in the large.

Exercise 3. If $q(P) \geqslant 0$, $q(P) \not\equiv 0$, in a closed space F, and if k is any positive integer, then if $L^k u \equiv 0$, we must have $u \equiv 0$, where $L = \Delta - q$, $L^k = L(L^{k-1})$.

Exercise 4. In a closed space F, let $u(P)$ be harmonic except at a finite number of points $Q_1,..., Q_k$, where

$$u(P) \sim r_i \gamma(P, Q_i),$$

and where $\gamma(P, Q)$ denotes the local fundamental singularity. Show that the sum $\sum r_i$ of the residues r_i must be zero, and give a 'physical interpretation' of the result.

Exercise 5. Show that the functions

$$(\operatorname{cosec} x)^{\frac{1}{2}} P_{\frac{1}{2}}^{l+\frac{1}{2}}(\cos x) P_l^m(\cos \theta) e^{im\varphi} \qquad (Rx = r)$$

are harmonic and single-valued in a three-dimensional space of constant positive curvature R. Here l and m are integers while P_l^m denotes the associated Legendre polynomial.

Exercise 6. Show that the functions

$$(\operatorname{cosech} x)^{\frac{1}{2}} P_{\frac{1}{2}}^{l+\frac{1}{2}}(\cosh x) P_l^m(\cos \theta) e^{im\varphi} \qquad (Rx = r)$$

are harmonic and single-valued in a three-dimensional space of constant negative curvature $-R$. Find the limits of these solutions as R tends to infinity.

5.6. The formulation of boundary value problems. Let D be a bounded domain in a Riemann space V_N, and let the bounding surface B be a twice differentiable surface of $N-1$ dimensions. We shall suppose that for our differential equation $\Delta u = qu$ there is known a fundamental singularity $\gamma(P, Q)$ defined throughout the entire space. This space V_N may be the double F of the domain D, or some other closed space of which D is a part; or it may be infinite and unbounded. We consider the problems of Dirichlet and Neumann both for the domain D and for the complementary or exterior domain $V_N - D$.

The method which we follow was invented by Poincaré and Fredholm. Let us attempt to construct a solution of the Dirichlet problem for D in the form of a double layer potential with an undetermined density $\mu(q)$:

$$W(P) = \int_B \mu(q) \frac{\partial \gamma(P, q)}{\partial n_q} \, dS_q. \tag{5.6.1}$$

Our task is now to choose the double layer density $\mu(q)$ in such a way that the limit $W_-(P)$ of $W(P)$, as P tends to the boundary from the interior, is equal to the assigned value $f(p)$. Since we require in effect that

$$W_-(p) = W(p) - \tfrac{1}{2}\mu(p) = f(p), \tag{5.6.2}$$

we see that the layer density $\mu(p)$ must satisfy the integral equation

$$f(p) = -\tfrac{1}{2}\mu(p) + \int_B \mu(q) \frac{\partial \gamma(p, q)}{\partial n_q} \, dS_q. \tag{5.6.3}$$

Similarly, if we wish to solve the Dirichlet problem for the region exterior to D, we may choose $\mu(p)$, if possible, so that

$$W_+(p) = f_1(p),$$

where the assigned exterior boundary values are denoted by $f_1(P)$. From (5.4.11), we have

$$W_+(p) = \tfrac{1}{2}\mu(p) + W(p), \tag{5.6.4}$$

and we are led to a similar integral equation

$$f_1(p) = \tfrac{1}{2}\mu(p) + \int_B \mu(q) \frac{\partial \gamma(p, q)}{\partial n_q} \, dS_q. \tag{5.6.5}$$

These two integral equations may be formally combined if we write $\varphi(p)$ as unknown function in place of $\tfrac{1}{2}\mu(p)$; and define

$$k(p, q) = 2 \frac{\partial \gamma(p, q)}{\partial n_q}. \tag{5.6.6}$$

We also introduce an eigenvalue parameter λ. Then the equation

$$f(p) = \varphi(p) - \lambda \int_B \varphi(q) k(p, q) \, dS_q \tag{5.6.7}$$

coincides with (5.6.3) when $\lambda = 1$, $f(p) = -f_1(p)$; and is identical with (5.6.5) when $\lambda = -1, f(p) = f_1(p)$. This equation (5.6.7) is known as Fredholm's integral equation with kernel $k(p, q)$. The success of this method of solving the various boundary value problems is due to Fredholm's method of solving (5.6.7), that is, of finding $\varphi(p)$ when $f(p)$ is given.

Analogously, let us treat the Neumann problem for D by constructing a suitable single layer potential,

$$V(P) = \int_B \sigma(q)\gamma(P,q)\, dS_q, \tag{5.6.8}$$

with an as yet undetermined single layer density $\sigma(q)$. In (5.4.13) we have a formula for the discontinuity of the normal derivative of $V(P)$. Let us choose $\sigma(q)$ so that, for the interior problem,

$$\frac{\partial V(p)}{\partial n_-} = \frac{\partial V(p)}{\partial n} + \tfrac{1}{2}\sigma(p) = F(p), \tag{5.6.9}$$

where the assigned data are the values of $F(p)$. Thus we find for $\sigma(p)$ the integral equation

$$F(p) = \tfrac{1}{2}\sigma(p) + \int_B \sigma(q)\frac{\partial\gamma(p,q)}{\partial n_p}\, dS_q. \tag{5.6.10}$$

For the exterior Neumann problem with data $F(p)$, we must set

$$\frac{\partial V(p)}{\partial n_+} = \frac{\partial V(p)}{\partial n} - \tfrac{1}{2}\sigma(p) = F(p), \tag{5.6.11}$$

and we likewise obtain for $\sigma(p)$ the equation

$$F(p) = -\tfrac{1}{2}\sigma(p) + \int_B \sigma(q)\frac{\partial\gamma(p,q)}{\partial n_p}\, dS_q. \tag{5.6.12}$$

Again, let us write $\psi(p)$ for $\tfrac{1}{2}\sigma(p)$, and make use of the definition (5.6.6). We find that the Fredholm equation

$$f(p) = \psi(p) - \lambda \int_B \psi(q)k(q,p)\, dS_q \tag{5.6.13}$$

agrees with (5.6.10) if $f(p) = F(p)$, $\lambda = -1$; and with (5.6.12) if $f(p) = -F(p)$, $\lambda = +1$. Comparing the two equations (5.6.7) and (5.6.13) we see that the kernels are the same except that the two argument points have been transposed. Indeed, (5.6.13) is said to be the transposed equation relative to (5.6.7). The next chapter, in which the solution of Fredholm's equation is derived, will show us that the properties of an integral equation and of its transpose are closely related.

To sum up, then, the interior Dirichlet problem and the exterior Neumann problem can be treated by means of the two Fredholm equations (5.6.7) and (5.6.13) with $\lambda = +1$. Similarly, the exterior Dirichlet problem and the interior Neumann problem may be solved via these same equations with $\lambda = -1$. It is a great advantage of this method that these four basic problems are reduced to closely related integral equations.

The kernel $k(p,q)$ is defined for all pairs of boundary points p and q except possibly when $p = q$, since then the function $\gamma(p,q)$ is infinite. Indeed, the kernel $k(p,q)$, being a derivative of $\gamma(p,q)$, also has a singularity when its two argument points coalesce. It will later be important to know that this singularity is not of too high an order. Let us examine $k(p,q)$ when $s = s(p,q)$ is small. We have

$$k(p,q) = 2\frac{\partial}{\partial n_q}\gamma(p,q) = \frac{2}{\omega_N(N-2)}\frac{\partial}{\partial n_q}(s^{-N+2}+\dots) \sim \frac{-2}{\omega_N}s^{-N+1}\frac{\partial s}{\partial n_q},$$

$$(5.6.14)$$

where terms of a lower order of infinity are omitted. Now let us suppose that the equation of the boundary B in this same neighbourhood is $\varphi(x^i) = 0$. Then the unit normal to B is $\nabla\varphi/\sqrt{\{(\nabla\varphi)^2\}}$; and

$$\frac{\partial s}{\partial n_q} = \frac{\nabla s . \nabla\varphi}{\sqrt{\{(\nabla\varphi)^2\}}} = \frac{1}{\sqrt{\{(\nabla\varphi)^2\}}}\frac{d\varphi}{ds}, \qquad (5.6.15)$$

where $s = s(p,q)$, and we have used (4.5.18). Now let x^r be a system of Riemannian coordinates at q; and let us write down the Taylor expansion about q for the function $\varphi(x^i)$. Thus, since we have assumed that $\varphi(x^i)$ is twice differentiable,

$$\varphi(p) = \varphi(q)+x^r\left(\frac{\partial\varphi}{\partial x^r}\right)_q+\tfrac{1}{2}x^rx^s\left(\frac{\partial^2\varphi}{\partial x^r\partial x^s}\right), \qquad (5.6.16)$$

where the second derivatives are evaluated at a suitable intermediate point. But p and q lie on the boundary, so $\varphi(p) = 0$, $\varphi(q) = 0$. Since $x^r = sp^r = sp_0^r$, we have now

$$s\left(\frac{d\varphi}{ds}\right)_q = sp_0^r\left(\frac{\partial\varphi}{\partial x^r}\right)_q = x^r\left(\frac{\partial\varphi}{\partial x^r}\right)_q = -\tfrac{1}{2}x^rx^s\left(\frac{\partial^2\varphi}{\partial x^r\partial x^s}\right)$$

$$= -\tfrac{1}{2}s^2p_0^rp_0^s\left(\frac{\partial^2\varphi}{\partial x^r\partial x^s}\right) = O(s^2). \qquad (5.6.17)$$

That is, $$\frac{d\varphi}{ds} = O(s),$$

so, as $s \to 0$, we see from (5.6.14), (5.6.15), and (5.6.17) that

$$k(p,q) \sim -\frac{2}{\omega_N}s^{1-N}\frac{\partial s}{\partial n_q}$$

$$= -\frac{2}{\omega_N}\frac{s^{1-N}}{\sqrt{\{(\nabla\varphi)^2\}}}\left(\frac{d\varphi}{ds}\right)_q \qquad (5.6.18)$$

$$= -\frac{2}{\omega_N}s^{1-N}O(s)$$

$$= O(s^{2-N}).$$

Thus the singularity of the kernel is of order $N-2$, so that the differentiation along the normal to the surface, which is indicated in (5.6.6), has not increased the order of the singularity. Since B is of $N-1$ dimensions, it is evident in particular that the integrals in (5.6.7) and (5.6.13) are convergent and indeed absolutely convergent. The importance of (5.6.18) will be evident when we return to the construction of the solutions for the Dirichlet and Neumann problems.

VI

LINEAR INTEGRAL EQUATIONS

THE preceding development of the theory of elliptic equations has led us, in two instances, to the study of integral equations of Fredholm's type. In § 5.3 we saw that the construction of a solution of Poisson's equation can be effected by solving an integral equation with domain of integration the region D in question. Also the boundary value problems were reduced in § 5.6 to the solution of Fredholm equations in which the domain of integration is the boundary surface B. The present chapter treats of the solution of these integral equations. In the first four sections are set forth the theorems of Fredholm which specify the conditions for solvability. In the next chapter we shall return to the differential equations point of view and will apply the Fredholm theory to the solution of the problems mentioned above.

The theory of integral equations constitutes a powerful and flexible tool for theoretical study of problems arising from differential equations, and is by no means restricted to these applications. In Chapter VIII we shall study eigenvalue problems which arise in connexion with eigenfunctions of the kind mentioned in § 5.1. As a preparation for this work we devote the last two sections of the present chapter to the study of integral equations with symmetric kernels—these are characteristic of eigenvalue problems arising from self-adjoint differential equations.

6.1. Fredholm's first theorem. The linear integral equation with fixed limits,

$$y(s) - \lambda \int_D k(s,t)y(t)\, dt = f(s), \tag{6.1.1}$$

was solved by Fredholm on the basis of an analogy with the theory of systems of linear algebraic equations, which arises when the integral is regarded as the limit of a sum. To describe this analogy, let us suppose that the domain D is an interval $a \leqslant t \leqslant b$ of the real line. Then let $s_1 = a,\ s_2,...,\ s_n = b$ be equally spaced points with interval δ, and let $k(s_i, s_j)$ be denoted by k_{ij}. The corresponding algebraic system is then

$$y(s_i) - \lambda \sum_{j=1}^{n} k_{ij} y(s_j)\delta = f(s_i) \qquad (i = 1,...,n).$$

A solution exists provided the determinant of this system is not zero; the determinant may be written

$$D_n(\lambda) = \begin{vmatrix} 1-\lambda\delta k_{11} & -\lambda\delta k_{12} & \cdot & \cdot & \cdot & \cdot & \cdot & \cdot \\ -\lambda\delta k_{21} & 1-\lambda\delta k_{22} & \cdot & \cdot & \cdot & \cdot & \cdot \\ \cdot & \cdot & \cdot & \cdot & \cdot & \cdot & \cdot & \cdot \\ \cdot & \cdot & \bullet & \cdot & \cdot & \cdot & \cdot & \cdot \\ \cdot & \cdot & \cdot & \cdot & \cdot & \cdot & \cdot & 1-\lambda\delta k_{nn} \end{vmatrix}$$

$$= 1 - \lambda\delta \sum_{j=1}^{n} k_{jj} + \frac{1}{2!}\lambda^2\delta^2 \sum_{ij} \begin{vmatrix} k_{ii} & k_{ij} \\ k_{ji} & k_{jj} \end{vmatrix} - \dots .$$

In the limit as $\delta \to 0$, $n \to \infty$, we should replace $\sum \delta$ by $\int \dots ds$; the determinant then becomes an infinite series whose $(r+1)$th term is an r-fold integral over the domain of integration. Also, if $D_n(s_i, s_j)$ denotes the cofactor of the term in $D_n(\lambda)$ which involves k_{ij}, we see that the solution of the algebraic system may be written

$$y(s_i) = \frac{\sum\limits_{i=1}^{n} D_n(s_i, s_k) f(s_k)\delta}{D_n(\lambda)} .$$

As $n \to \infty$, this formula becomes, in the limit, an integral over the values of the function $f(s)$.

Returning to the integral equation (6.1.1), let us define the functions which, in this analogy, correspond to the determinant and to its minors. We shall assume that the domain D is a bounded region of N dimensions, the volume $V = \int_D dV$ being finite. Also, let the 'kernel' $k(s, t)$ be a bounded continuous function of its arguments s and t which represent points ranging over D. We shall, for convenience, omit the symbol D from the integrals, so that the integral sign alone will, for the rest of this chapter, denote the integral over the given domain D.

The analogy just described suggests the following definition of the 'determinant of the coefficients' for the integral equation. Let $s_1, \dots, s_n, t_1, \dots, t_n$ be variables of integration ranging over D, and define the determinant symbol

$$k\begin{pmatrix} s_1, \dots, s_n \\ t_1, \dots, t_n \end{pmatrix} = \begin{vmatrix} k(s_1, t_1) & k(s_1, t_2) & \cdot & \cdot & \cdot & k(s_1, t_n) \\ \cdot & \cdot & \cdot & \cdot & \cdot & \cdot & \cdot \\ \cdot & \cdot & \cdot & \cdot & \cdot & \cdot & \cdot \\ \cdot & \cdot & \cdot & \cdot & \cdot & \cdot & \cdot \\ k(s_n, t_1) & \cdot & \cdot & \cdot & \cdot & k(s_n, t_n) \end{vmatrix}. \quad (6.1.2)$$

Furthermore let $A_0 = 1$, and

$$A_n = \frac{1}{n!} \int \cdots \int k\binom{s_1,\ldots,s_n}{s_1,\ldots,s_n} ds_1 \ldots ds_n. \tag{6.1.3}$$

Then the determinant $D(\lambda)$ of Fredholm is given by the power series

$$D(\lambda) = \sum_{n=0}^{\infty} A_n(-\lambda)^n. \tag{6.1.4}$$

The merit of Fredholm's approach over the successive approximation method, which latter is useful only if $|\lambda|$ is sufficiently small, is that the determinant $D(\lambda)$ is an entire function of the complex variable λ and so is defined and finite for all values of λ.

To demonstrate this property of the series (6.1.4), we must estimate the magnitude of the coefficients A_n, and therefore of the determinant (6.1.2). Let us observe that in a Euclidean n-dimensional space, this determinant is the volume of a certain parallelepiped, which has one vertex at the origin O, and edges joining the origin to the n points P_i ($i = 1,\ldots,n$), with Cartesian coordinates $k(s_i, s_j)$ ($j = 1,\ldots,n$). If its sides are of given lengths, this volume is greatest when its shape is rectangular. Now let M be a bound for the absolute value of $k(s,t)$ in D. Then the distances OP_i are

$$d_i = \sqrt{\left\{ \sum_{j=1}^{n} k(s_i,s_j)^2 \right\}} \leqslant \sqrt{\left\{ \sum_{j=1}^{n} M^2 \right\}} \leqslant M\sqrt{n}.$$

Therefore, finally, the volume of the parallelepiped does not exceed $M^n n^{\frac{1}{2}n}$ which is the volume of the n-cube with side of length $M\sqrt{n}$. Therefore,

$$\left| k\binom{s_1,\ldots,s_n}{t_1,\ldots,t_n} \right| \leqslant M^n n^{\frac{1}{2}n}.$$

Thus

$$|A_n| \leqslant \frac{1}{n!} n^{\frac{1}{2}n}(MV)^n,$$

and the series (6.1.4) is dominated by the series

$$\sum_{n=0}^{\infty} \frac{1}{n!} n^{\frac{1}{2}n}(MV|\lambda|)^n,$$

of which successive terms have the ratio

$$\frac{MV|\lambda|}{n+1} \bigg/ \sqrt{\left\{ \frac{(n+1)^{n+1}}{n^n} \right\}} = \frac{MV|\lambda|}{(n+1)^{\frac{1}{2}}} \bigg/ \sqrt{\left\{ \left(1+\frac{1}{n}\right)^n \right\}} < \frac{MV|\lambda|}{(n+1)^{\frac{1}{2}}} e.$$

Thus the ratio of successive terms tends to zero for all values of $|\lambda|$, and it follows that the series for $D(\lambda)$ converges in the entire λ-plane and defines $D(\lambda)$ as an entire or integral function of the complex variable λ.

We must also define certain related functions known as the 'first minors' of $D(\lambda)$. Let $A_0(s,t) = k(s,t)$ and let

$$A_n(s,t) = \frac{1}{n!} \int \cdots \int k \begin{pmatrix} s & s_1,\ldots,s_n \\ t & s_1,\ldots,s_n \end{pmatrix} ds_1 \ldots ds_n. \qquad (6.1.5)$$

Then we define the first minor

$$D(s,t,\lambda) = \sum_{n=0}^{\infty} A_n(s,t)(-\lambda)^n. \qquad (6.1.6)$$

A very similar convergence proof shows that $D(s,t,\lambda)$ is also an entire function of the complex variable λ.

Now let us expand the determinant in the integrand of (6.1.5) by elements of its first row. There results a term multiplied by $k(s,t)$ and n further terms with factors $k(s,s_i)$ $(i = 1,\ldots,n)$, viz.,

$$k(s,\,t)k \begin{pmatrix} s_1,\ldots,s_n \\ s_1,\ldots,s_n \end{pmatrix} + \sum_{\mu=1}^{n} k(s,s_\mu)k \begin{pmatrix} s_1, & \ldots, & s_n \\ s_1,\ldots,s_{\mu-1}, & t, & s_{\mu+1},\ldots,s_n \end{pmatrix}.$$

When the integrations in (6.1.5) have been carried out as far as possible, and (6.1.3) used, we find that the n terms in the summation above all yield the same result, and that

$$A_n(s,t) = A_n k(s,t) - \int k(s,\sigma)A_{n-1}(\sigma,t)\, d\sigma. \qquad (6.1.7)$$

This formula is true for $n = 0$ if we assume $A_{-1}(s,t) \equiv 0$. Multiply (6.1.7) by $(-\lambda)^n$ and sum over n from zero to infinity; there results the formula

$$D(s,t,\lambda) = k(s,t)D(\lambda) + \lambda \int k(s,\sigma)D(\sigma,t,\lambda)\, d\sigma. \qquad (6.1.8)$$

We shall now define the resolvent kernel

$$R(s,t,\lambda) = \frac{D(s,t,\lambda)}{D(\lambda)}, \qquad (6.1.9)$$

which will appear in the final solution formula. It is clear that the resolvent is an analytic function of λ, and that its only singularities are poles which must occur at the zeros of the denominator $D(\lambda)$. From (6.1.8) and (6.1.9) there follows the resolvent equation

$$R(s,t,\lambda) - k(s,t) = \lambda \int k(s,\sigma)R(\sigma,t,\lambda)\, d\sigma. \qquad (6.1.10)$$

If the derivation of (6.1.10) is repeated, with the modification that the determinant expansion should be by columns, rather than by rows, a second resolvent equation, the 'transpose' of (6.1.10) is found, namely,

$$R(s,t,\lambda) - k(s,t) = \lambda \int R(s,\sigma,\lambda)k(\sigma,t)\, d\sigma. \qquad (6.1.11)$$

These two formulae enable us to solve the Fredholm equation when $D(\lambda)$ is different from zero, so that $R(s, t, \lambda)$ is regular.

We claim that the solution is then unique, and is given by

$$y(s) = f(s) + \lambda \int R(s, t, \lambda) f(t) \, dt. \qquad (6.1.12)$$

To prove these assertions, let us suppose that $y(s)$ is any solution of (6.1.1). Then, substituting from (6.1.1), we have

$$f(s) + \lambda \int R(s, t, \lambda) f(t) \, dt$$

$$= y(s) - \lambda \int k(s, t) y(t) \, dt + \lambda \int R(s, t, \lambda) \Big[y(t) - \lambda \int k(t, \sigma) y(\sigma) \, d\sigma \Big] \, dt$$

$$= y(s) + \lambda \int y(t) \, dt \Big[R(s, t, \lambda) - k(s, t) - \lambda \int R(s, \sigma, \lambda) k(\sigma, t) \, d\sigma \Big]$$

$$= y(s),$$

in view of the resolvent formula (6.1.11). Thus any solution must be of the form (6.1.12). Conversely, (6.1.12) is a solution, because, on substitution of (6.1.12) into (6.1.1) we find

$$f(s) + \lambda \int R(s, t, \lambda) f(t) \, dt - \lambda \int k(s, t) f(t) \, dt - \lambda^2 \iint R(t, \sigma, \lambda) k(s, t) f(\sigma) \, d\sigma dt$$

$$= f(s) + \lambda \int f(t) \, dt \Big[R(s, t, \lambda) - k(s, t) - \lambda \int k(s, \sigma) R(\sigma, t, \lambda) \, d\sigma \Big]$$

$$= f(s),$$

in view of the first resolvent relation (6.1.10). Thus the solution of the integral equation exists and is unique provided that $D(\lambda)$ is not zero. In particular, if $f(s) \equiv 0$, the solution (6.1.12) is identically zero, so that the homogeneous equation corresponding to (6.1.1) has no (continuous) solution not identically zero. To sum up, we have Fredholm's first theorem: *If $D(\lambda) \neq 0$, equation (6.1.1) has the unique solution (6.1.12).*

Associated with (6.1.1) is the transposed equation

$$z(s) - \lambda \int z(t) k(t, s) \, dt = f(s) \qquad (6.1.13)$$

in which the kernel $k(t, s)$ is the transposed coefficient matrix. The determinant $D(\lambda)$ defined by (6.1.4) is clearly also the determinant of the transposed kernel, and the first minor is now $D(t, s, \lambda)$. Thus the resolvent is $R(t, s, \lambda)$, and the solution formula which corresponds to (6.1.12) is easily written down.

6.2. Fredholm's second theorem. The first theorem of Fredholm does not apply if λ is a root of the equation $D(\lambda) = 0$. By analogy with the properties of linear algebraic equations, it might be expected that the homogeneous equation

$$y(s) - \lambda \int k(s,t)y(t)\, dt = 0 \qquad (6.2.1)$$

would then have a solution not identically zero. Such a value of the parameter λ is known as an eigenvalue, or characteristic value; and the corresponding solution function is called an eigenfunction. If $D(\lambda) = 0$ and if $D(s,t,\lambda)$ is not identically zero, we see from (6.1.8) that $D(s,t,\lambda)$, as a function of s, is a solution of (6.2.1). However, $D(s,t,\lambda)$ may be zero identically.

In order to treat the question in its general form, we must define minor determinants of higher order than the first. Let

$$A_n\!\begin{pmatrix} s_1,\dots, s_m \\ t_1,\dots, t_m \end{pmatrix} = \frac{1}{n!}\int \dots \int k\begin{pmatrix} s_1,\dots, s_m,\, \sigma_1,\dots,\sigma_n \\ t_1,\dots, t_m,\, \sigma_1,\dots,\sigma_n \end{pmatrix} d\sigma_1 \dots d\sigma_n, \qquad (6.2.2)$$

and, analogously to (6.1.6), let us define the mth minor

$$D\!\begin{pmatrix} s_1,\dots, s_m \\ t_1,\dots, t_m \end{pmatrix}\lambda\Bigg) = \sum_{n=0}^{\infty} A_n\!\begin{pmatrix} s_1,\dots, s_m \\ t_1,\dots, t_m \end{pmatrix}(-\lambda)^n. \qquad (6.2.3)$$

As before, this series converges for all values of λ and so this mth minor is an entire function of λ. According to this notation, we might have written $D(^s_t\lambda)\cdot$instead of $D(s,t,\lambda)$; but the latter seems more convenient.

From (6.1.3) and (6.2.2) we see that

$$\int \dots \int A_n\!\begin{pmatrix} s_1,\dots, s_m \\ s_1,\dots, s_m \end{pmatrix} ds_1 \dots ds_m = \frac{(m+n)!}{n!} A_{m+n},$$

and, upon multiplication by $(-\lambda)^n$ and summation over n from zero to infinity, we obtain

$$\int \dots \int D\!\begin{pmatrix} s_1,\dots, s_m \\ s_1,\dots, s_m \end{pmatrix}\lambda\Bigg) ds_1\dots ds_m = (-1)^m \frac{d^m}{d\lambda^m} D(\lambda). \qquad (6.2.4)$$

Let us find the relations among the minors which correspond to the resolvent formulae (6.1.10) and (6.1.11). Expansion of the determinant in the integrand of (6.2.2) by elements of the first row gives rise to m terms containing a factor $k(s_1, t_i)$ $(i = 1,\dots, m)$, and n further terms with a factor $k(s_1, \sigma_j)$ $(j = 1,\dots, n)$. If now the integrations in (6.2.2) are performed, as far as possible, these latter n terms all yield the same result, and we find

$$A_n\begin{pmatrix}s_1,...,s_m\\t_1,...,t_m\end{pmatrix} = \sum_{\mu=1}^{m}(-1)^\mu k(s_1,t_\mu)A_n\begin{pmatrix}s_2,&...,&s_m\\t_1,...,t_{\mu-1},t_{\mu+1},...,t_m\end{pmatrix} -$$

$$- \int k(s_1,\sigma)A_{n-1}\begin{pmatrix}\sigma,s_2,...,s_m\\t_1,t_2,...,t_m\end{pmatrix} d\sigma.$$

Multiplying by $(-\lambda)^n$ and summing, we have in view of (6.2.3),

$$D\begin{pmatrix}s_1,...,s_m\\t_1,...,t_m\end{pmatrix}\lambda = \sum_{\mu=1}^{m}(-1)^\mu k(s_1,t_\mu)D\begin{pmatrix}s_2,&...,&s_m\\t_1,...,t_{\mu-1},t_{\mu+1},...,t_m\end{pmatrix}\lambda +$$

$$+\lambda \int k(s_1,\sigma)D\begin{pmatrix}\sigma,s_2,...,s_m\\t_1,t_2,...,t_m\end{pmatrix}\lambda\, d\sigma. \qquad (6.2.5)$$

This identity generalizes (6.1.8). Expansion by the elements of any other row in (6.2.2) leads to a similar identity. There is also an identity to be found by expansion by the elements of the first column; it is

$$D\begin{pmatrix}s_1,...,s_m\\t_1,...,t_m\end{pmatrix}\lambda = \sum_{\mu=1}^{m}(-1)^\mu k(s_\mu,t_1)D\begin{pmatrix}s_1,...,s_{\mu-1},s_{\mu+1},...,s_m\\t_2,&...,&t_m\end{pmatrix}\lambda +$$

$$+\lambda \int k(\sigma,t_1)D\begin{pmatrix}s_1,...,s_m\\\sigma,t_2,...,t_m\end{pmatrix}\lambda\, d\sigma. \qquad (6.2.6)$$

These two formulae will play the role of the resolvent formulae of the preceding section.

Let us now suppose that $D(\lambda)$, together with the minors (6.2.3) up to but not including those of mth order, are identically zero for the value of λ in question. It follows from (6.2.4) that $D(\lambda)$ has a zero of order at least m at this value of λ. Since the mth minor is assumed not to vanish identically, there are values of the s_i and t_i such that

$$D\begin{pmatrix}s_1,...,s_m\\t_1,...,t_m\end{pmatrix}\lambda \neq 0. \qquad (6.2.7)$$

Now, by hypothesis, the terms in the sum on the right in (6.2.5) vanish, and we see that the mth minor, as a function of $s = s_1$, is a solution of the homogeneous equation which, by (6.2.7), is not identically zero. Now any one of the m variables $s_1,...,s_m$ might be singled out in this way. It follows that each of the m functions

$$y_k(s) = \frac{D\begin{pmatrix}s_1,...,s_{k-1},s,s_{k+1},...,s_m\\t_1,&...,&t_m\end{pmatrix}\lambda}{D\begin{pmatrix}s_1,&...,&s_m\\t_1,&...,&t_m\end{pmatrix}\lambda} \qquad (6.2.8)$$

is a solution of the homogeneous equation (6.2.1). Actually these m solutions are linearly independent, for the following reason. If in (6.1.2) we set two of the argument points s_i equal, the determinant vanishes. Thus in (6.2.8) we see that $y_k(s_i) = 0$, for $i \neq k$, whereas $y_k(s_k) = 1$. (It is clear that the numbers s_i in (6.2.7) must be distinct). If now there exists a relation

$$\sum_k c_k y_k(s) \equiv 0,$$

we may set $s = s_i$, and it follows that $c_i = 0$. That is, the $y_k(s)$ are linearly independent as stated.

Conversely, we can, with the aid of (6.2.5), show that any solution of (6.2.1) must be a linear combination of these m functions. Let us first define a kernel which corresponds to the resolvent function of the preceding section, namely,

$$\Gamma(s,t,\lambda) = \frac{D\begin{pmatrix} s, s_1, \ldots, s_m \\ t, t_1, \ldots, t_m \end{pmatrix} \lambda}{D\begin{pmatrix} s_1, \ldots, s_m \\ t_1, \ldots, t_m \end{pmatrix} \lambda}, \tag{6.2.9}$$

where the s_i and t_i are chosen as in (6.2.7). Now write down (6.2.5) with m replaced by $m+1$, and with the extra arguments s and t. Dividing the equation so obtained by the quantity in (6.2.7) and taking note of (6.2.9), we find

$$\Gamma(s,t,\lambda) - k(s,t) - \lambda \int \Gamma(s,\sigma,\lambda)k(\sigma,t)\,d\sigma = -\sum_{\mu=1}^{m} k(s_\mu,t)y_\mu(s). \tag{6.2.10}$$

If now $y(t)$ is any solution of (6.2.1), let us multiply (6.2.10) by $y(t)$ and integrate over t. We find, using (6.2.1) in all terms but the first,

$$\int y(t)\Gamma(s,t,\lambda)\,dt - \frac{y(s)}{\lambda} - \int y(\sigma)\Gamma(s,\sigma,\lambda)\,d\sigma = -\sum_{\mu=1}^{m} \frac{y(s_\mu)}{\lambda} y_\mu(s).$$

Since the two integrals cancel we see that

$$y(s) = \sum_{\mu=1}^{m} y(s_\mu)y_\mu(s) \tag{6.2.11}$$

is indeed a linear sum of the m independent eigenfunctions $y_k(s)$.

By the same token, the homogeneous transposed equation

$$z(s) - \lambda \int k(t,s)z(t)\,dt = 0 \tag{6.2.12}$$

has the same number m of linearly independent solutions. Thus we have established Fredholm's second theorem: *If $D(\lambda)$ has a zero of order n, the homogeneous equation (6.2.1) and its transpose (6.2.12) have the same number m ($1 \leqslant m \leqslant n$) of linearly independent solutions.*

Thus if $D(\lambda)$ vanishes there exists at least one solution of the homogeneous equation (6.2.1), and therefore a solution of the non-homogeneous equation (6.1.1) cannot be unique. Since $D(\lambda)$ either is or is not zero, we conclude from these two theorems the following 'Fredholm alternative': *either the non-homogeneous equation has always a unique solution or the homogeneous equation has a non-zero solution.* Thus, if we can show that any solution of an integral equation is necessarily unique, then the first of the two alternatives holds, and a solution must exist. From uniqueness we may infer existence.

Exercise 1. Show that not every one of the Fredholm minors (6.2.3) can vanish identically for any given value of λ, $m = 0, 1, 2,...$. *Hint*: Use (6.2.4).

Exercise 2. If $y_k(s,\lambda)$, $z_k(s,\mu)$ are eigenfunctions of (6.2.1) and (6.2.12) with parameters $\mu \neq \lambda$, show that

$$\int y_k(s,\lambda)z_k(s,\mu)\, ds = 0.$$

6.3. Fredholm's third theorem. The non-homogeneous Fredholm equation

$$y(s) - \lambda \int k(s,t)y(t)\, dt = f(s) \tag{6.3.1}$$

has been shown to possess a unique solution if $D(\lambda) \neq 0$. We shall now suppose that $D(\lambda) = 0$, and investigate this question: when does (6.3.1) have a solution if $f(s)$ is not identically zero? Since systems of linear algebraic equations with vanishing coefficient determinant may be incompatible, it is to be expected that (6.3.1) may also fail to possess a solution.

Let m be the order of the lowest non-vanishing minor of $D(\lambda)$, and let $z_k(s)$ $(k = 1,...,m)$, be the m eigenfunctions of the homogeneous transposed equation (6.2.12). Then, if a solution function $y(s)$ of (6.3.1) exists, we have

$$\int f(s)z_k(s)\, ds = \int y(s)z_k(s)\, ds - \lambda \int\int k(s,t)y(t)z_k(s)\, dsdt$$
$$= \int y(s)\, ds\Big[z_k(s) - \lambda \int k(t,s)z_k(t)\, dt\Big] = 0. \tag{6.3.2}$$

The bracketed term is zero in view of (6.2.12). We see, therefore, that a necessary condition for the solvability of (6.3.1) is that the non-homogeneous term $f(s)$ be orthogonal to each solution of the homogeneous transposed equation (6.2.12).

Conversely, this condition of orthogonality is sufficient in order that a solution exist. Indeed, let us suppose that the condition (6.3.2) is satisfied; then the function

$$y_0(s) = f(s) + \lambda \int \Gamma(s,t,\lambda)f(t)\, dt \tag{6.3.3}$$

is a solution. Here $\Gamma(s, t, \lambda)$ is defined by (6.2.9). To verify this assertion, we substitute (6.3.3) in the left-hand side of (6.3.1), obtaining

$$f(s) + \lambda \int \Gamma(s, t, \lambda) f(t)\, dt - \lambda \int k(s, t)\Big[f(t) + \lambda \int \Gamma(t, \sigma, \lambda) f(\sigma)\, d\sigma\Big]\, dt$$
$$= f(s) + \lambda \int f(t)\, dt \Big[\Gamma(s, t, \lambda) - k(s, t) - \lambda \int k(s, \sigma)\Gamma(\sigma, t, \lambda)\, d\sigma\Big]. \tag{6.3.4}$$

Now the bracketed term may be transformed as follows. In (6.2.6) write $m + 1$ for m, and introduce the additional arguments s and t. Rearranging, dividing by the mth minor (6.2.7) and noting (6.2.9), we obtain an equation which is the 'transpose' of (6.2.10), namely

$$\Gamma(s, t, \lambda) - k(s, t) - \lambda \int k(s, \sigma)\Gamma(\sigma, t, \lambda)\, d\sigma = - \sum_{\mu=1}^{m} k(s, t_\mu) z_\mu(t). \tag{6.3.5}$$

By (6.3.4), we find the left-hand side of (6.3.1) now becomes

$$f(s) - \lambda \sum_{\mu=1}^{m} k(s, t_\mu) \int f(t) z_\mu(t)\, dt = f(s), \tag{6.3.6}$$

since the orthogonality condition (6.3.2) is assumed to hold. Thus (6.3.3) is indeed a solution.

Since the difference of any two solutions of (6.3.1) is a solution of the homogeneous equation, the most general solution of (6.3.1) is

$$y_0(s) + \sum_{\mu=1}^{m} c_\mu y_\mu(s). \tag{6.3.7}$$

To sum up, we have established Fredholm's third theorem: *The non-homogeneous integral equation*

$$y(s) - \lambda \int k(s, t)y(t)\, dt = f(s) \tag{6.3.1}$$

has a solution if and only if

$$\int f(s) z_\mu(s)\, ds = 0, \tag{6.3.8}$$

for all solutions $z_\mu(s)$ of the homogeneous transposed equation

$$z(s) - \lambda \int k(s, t)z(t)\, dt = 0. \tag{6.3.9}$$

The most general solution has the form (6.3.7).

The three Fredholm theorems characterize completely the behaviour of the integral equation. It may be noted that the third theorem includes the first as a special case, namely, that in which the homogeneous transposed equation has no solutions.

Exercise. Find a solution of the Fredholm equation in the form of a power series

$$y(s) = \sum_{n=0}^{\infty} u_n(s)\lambda^n \quad (u_0(s) = f(s)),$$

by substituting in the equation and calculating the $u_n(s)$ successively by equating powers of λ. Show that the series converges for $|\lambda|MV < 1$.

6.4. Iterated kernels. The boundary value problems formulated in the preceding chapter were reduced to Fredholm equations in which the kernel $k(s, t)$ has a singularity when its two argument points coalesce. We must therefore extend the Fredholm theorems to such equations with singular kernels. This can be done by means of certain iterated kernels related to the given kernel. In order to define these, we shall find it convenient to use an operator notation for writing the integral equation. Thus, the integral operator with kernel $k(s, t)$ will be denoted by K:

$$Ky(s) \equiv \int k(s, t)y(t)\, dt, \tag{6.4.1}$$

and its transpose by $\tilde{K}$:

$$\tilde{K}y(s) \equiv \int k(t, s)y(t)\, dt. \tag{6.4.2}$$

The Fredholm equation (6.1.1) is then

$$y(s) - \lambda Ky(s) = f(s), \tag{6.4.3}$$

and it may also be written, by iteration,

$$y(s) = f(s) + \lambda Ky(s)$$
$$= f(s) + \lambda Kf(s) + \lambda^2 K^2 y(s)$$
$$= \quad . \quad . \quad . \quad . \quad . \quad .$$
$$= f(s) + \lambda Kf(s) + \ldots + \lambda^{n-1}K^{n-1}f(s) + \lambda^n K^n y(s). \tag{6.4.4}$$

Here $K^2 y(s)$ denotes $KKy(s)$, and has the kernel

$$k^{(2)}(s, t) = \int k(s, \sigma)k(\sigma, t)\, d\sigma.$$

The pth 'power' operator K^p has as kernel the pth iterated kernel of $k(s, t)$, namely,

$$k^{(p)}(s, t) = \int k^{(p-1)}(s, \sigma)k(\sigma, t)\, d\sigma. \tag{6.4.5}$$

When the domain D is of N dimensions, the singularity of the kernels we shall need to consider is of the type

$$k(s, t) \sim cr^{-N+\alpha} \quad (0 < \alpha < N),$$

where $r = r(s, t)$ denotes the distance function. Then the iterated kernels are defined by integrals of the type

$$I(s, t) = \int r(s, \sigma)^{-N+\alpha} \, r(\sigma, t)^{-N+\beta} \, d\sigma,$$

where again $0 < \alpha < N$, $0 < \beta < N$. We will now show that when $s \to t$, the singularity of $I(s, t)$ is of order $r^{-N+\alpha+\beta}$. Let s, t be fixed points of D and let $2\epsilon = r(s, t)$. Then we may divide D into four parts as follows (Fig. 6):

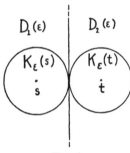

FIG. 6.

(1) the sphere $K_\epsilon(s)$ defined by $r(s, \sigma) \leqslant \epsilon$,
(2) the region $D_1(\epsilon)$ in which $r(s, \sigma) < r(\sigma, t)$, less $K_\epsilon(s)$,
(3) the sphere $K_\epsilon(t)$ defined by $r(\sigma, t) \leqslant \epsilon$,
(4) the region $D_2(\epsilon)$ in which $r(\sigma, t) < r(s, \sigma)$, less $K_\epsilon(t)$.

The corresponding integrals may be estimated as follows:

$$I_1 = \int_{K_\epsilon(s)} r(s, \sigma)^{-N+\alpha} \, r(\sigma, t)^{-N+\beta} \, d\sigma$$

$$\leqslant \epsilon^{-N+\beta} \int_{K_\epsilon(s)} r(s, \sigma)^{-N+\alpha} \, r^{N-1} \, dr d\Omega$$

$$\leqslant K_1 \epsilon^{-N+\beta} \int_0^\epsilon r^{\alpha-1} \, dr = \frac{K_1}{\alpha} \epsilon^{-N+\alpha+\beta},$$

where K_1 is a constant independent of the point s. Also,

$$I_2 = \int_{D_1(\epsilon)} r(s, \sigma)^{-N+\alpha} \, r(\sigma, t)^{-N+\beta} \, d\sigma$$

$$\leqslant \int_{D_1(\epsilon)} r(s, \sigma)^{-2N+\alpha+\beta} \, r^{N-1} \, dr d\Omega$$

$$\leqslant \int_{D-K_\epsilon(s)} r^{-N+\alpha+\beta-1} \, dr d\Omega$$

$$\leqslant K_1 \int_\epsilon^\infty r^{-N+\alpha+\beta-1} \, dr \leqslant \frac{K_1 \epsilon^{-N+\alpha+\beta}}{N-\alpha-\beta}.$$

The estimations of I_3 and I_4 are quite similar, and lead to bounds of the same order $\epsilon^{-N+\alpha+\beta}$ as $\epsilon \to 0$. Thus, finally, if $\alpha+\beta < N$,

$$I(s,t) \leqslant K\epsilon^{-N+\alpha+\beta} \leqslant \bar{K}r(s,t)^{-N+\alpha+\beta},$$

where $\bar{K}$ is a constant independent of s and t.

Therefore, if (as in § 5.6, with $N-1$ instead of N) we have $\alpha = 1$, then by successive application of this estimate we see that

$$k^{(p)}(s,t) = O\{r(s,t)^{-N+p}\} \quad (p = 1,2,...,N-1), \tag{6.4.6}$$

and so the iterated kernels of order N or higher are continuous for $s = t$. Thus the effect of iteration on this singularity is to reduce the order by one degree at a time. We shall now study the Fredholm theory for equations, such that the Nth iterated kernel $k^{(N)}(s,t)$ is continuous.

From (6.4.3) and (6.4.4) we see that if $y(s)$ is a solution of

$$y(s) - \lambda Ky(s) = f(s), \tag{6.4.1}$$

then $y(s)$ is also a solution of the equation with a continuous kernel

$$y(s) - \lambda^N K^N y(s) = g(s), \tag{6.4.7}$$

where

$$g(s) = f(s) + \lambda Kf(s) + ... + \lambda^{N-1}K^{N-1}f(s). \tag{6.4.8}$$

Suppose now that the determinant $D_N(\lambda^N)$ of (6.4.7) is not zero, so that a solution of (6.4.7) exists and is unique. Let $y(s)$ be this solution, then if we define

$$y_1(s) = f(s) + \lambda Ky(s), \tag{6.4.9}$$

we have from (6.4.7)

$$\begin{aligned}
y_1(s) &= f(s) + \lambda K[g(s) + \lambda^N K^N y(s)] \\
&= f(s) + \lambda Kf(s) + \lambda^2 K^2 f(s) + ... + \lambda^N K^N f(s) + \lambda^{N+1}K^{N+1}y(s) \\
&= g(s) + \lambda^N K^N[f(s) + \lambda Ky(s)] \\
&= g(s) + \lambda^N K^N y_1(s).
\end{aligned} \tag{6.4.10}$$

Thus $y_1(s)$ is also a solution of (6.4.7), and must therefore be equal to $y(s)$ under our hypothesis of uniqueness for the solution of (6.4.7). That is,

$$y_1(s) = f(s) + \lambda Ky(s) = y(s). \tag{6.4.11}$$

This is exactly a statement that $y(s)$ is a solution of (6.4.1). Thus, if the iterated equation has a unique solution, that solution satisfies the original equation. From (6.4.1), (6.4.7), and (6.4.8) it is easily seen that (6.4.1) can then have at most one solution, namely, $y(s)$.

Let us now consider the homogeneous equation

$$y(s) - \lambda K y(s) = 0, \qquad (6.4.12)$$

and its iterated equation

$$y(s) - \lambda^N K^N y(s) = 0. \qquad (6.4.13)$$

Let $y_i(s)$ $(i = 1,...,m)$ be the finite number of linearly independent solutions of (6.4.13)—the eigenfunctions. Then

$$\lambda K y_i(s) = \lambda^{N+1} K^{N+1} y_i(s) = \lambda^N K^N \{\lambda K y_i(s)\}, \qquad (6.4.14)$$

so that the m functions $\lambda K y_i(s)$ are also eigenfunctions. These latter must therefore be linear combinations of the $y_i(s)$:

$$\lambda K y_i(s) = \sum_j c_{ij} y_j(s), \qquad (6.4.15)$$

where the coefficients c_{ij} are m^2 constants. Regarding these constants as coefficients of an $m \times m$ matrix C, we may write (6.4.15) in the form

$$\lambda K \bar{y}(s) = C \bar{y}(s),$$

where $\bar{y}(s)$ is a vector with components $y_i(s)$ $(i = 1,...,m)$. Thus

$$\lambda^2 K^2 \bar{y}(s) = C^2 \bar{y}(s), \quad ..., \quad \lambda^p K^p \bar{y}(s) = C^p \bar{y}(s),$$

and

$$\bar{y}(s) = \lambda^N K^N \bar{y}(s) = C^N \bar{y}(s), \qquad (6.4.16)$$

since the $y_i(s)$ are eigenfunctions of (6.4.13). From (6.4.16) we see that the matrix C satisfies the matrix equation

$$C^N - I = 0.$$

It follows that the characteristic roots ρ_n of the matrix C, which are roots of the polynomial equation

$$|C - \rho I| \equiv \det |C - \rho I| = 0,$$

satisfy $\rho_n^N - 1 = 0$; that is, they are Nth roots of unity. Again, according to the theory of matrices, we may perform a rotation of axes in the m-dimensional vector space of the eigenfunctions $y_i(s)$, which brings to principal axis form the bilinear expression

$$\sum c_{ij} x^i x^j.$$

We then obtain m new eigenfunctions, linear combinations of the earlier ones, which we shall again denote by $y_i(s)$ $(i = 1,...,m)$, and which now satisfy equations

$$\lambda K y_i(s) = \rho_i y_i(s), \qquad (6.4.17)$$

in place of (6.4.15). Here ρ_i is an Nth root of unity. For those values of i, if any, for which $\rho_i = 1$, we see that $y_i(s)$ is an eigenfunction of (6.4.1);

otherwise, $y_i(s)$ is an eigenfunction of the same kernel $k(s,t)$ but with a different eigenvalue. By the second Fredholm theorem, the transpose

$$z(s)-\lambda^N \tilde{K}^N z(s) = 0 \qquad (6.4.18)$$

has the same number m of eigenfunctions $z_i(s)$. As before, these can be chosen in such a way that each satisfies an equation

$$\lambda \tilde{K} z_i(s) = \rho_i z_i(s), \qquad (6.4.19)$$

where ρ_i is again some Nth root of unity.

Let us now consider the solvability of the non-homogeneous equation

$$y(s)-\lambda K y(s) = f(s), \qquad (6.4.1)$$

when the solution of the iterated equation (6.4.7) is not necessarily unique. We shall establish Fredholm's third theorem for (6.4.1). It is clear that the condition of orthogonality of $f(s)$ to all solutions $\bar{z}_i(s)$ of

$$\bar{z}_i(s)-\lambda \tilde{K} \bar{z}_i(s) = 0 \qquad (6.4.20)$$

is again necessary. Suppose then that

$$\int f(s)\bar{z}_i(s)\, ds = 0 \qquad (i = 1,...,m). \qquad (6.4.21)$$

Now the iterated equation (6.4.7) has a solution if and only if $g(s)$, defined by (6.4.8), is orthogonal to all solutions $z_i(s)$ ($i = 1,..., m$), of (6.4.18). In view of (6.4.19), we have

$$\int g(s)z_i(s)\, ds = \sum_{j=0}^{N-1} \int \lambda^j K^j f(s)z_i(s)\, ds$$

$$= \sum_{j=0}^{N-1} \lambda^j \int f(s)\tilde{K}^j z_i(s)\, ds$$

$$= \sum_{j=0}^{N-1} \rho_i^j \int f(s)z_i(s)\, ds. \qquad (6.4.22)$$

Now, if $z_i(s)$ satisfies (6.4.20), that is, if $\rho_i = 1$, then from (6.4.21) we see that the integral on the right vanishes. Also, if $\rho_i \neq 1$,

$$\sum_{j=0}^{N-1} \rho_i^j = \frac{1-\rho_i^N}{1-\rho_i} = 0.$$

since ρ_i is an Nth root of unity. Hence in both cases,

$$\int g(s)z_i(s)\, ds = 0, \qquad (6.4.23)$$

and it follows that (6.4.7) has a solution, the most general form of which is

$$y(s) = y_0(s)+\sum_{i=1}^{m} a_i y_i(s), \qquad (6.4.24)$$

where the a_i are arbitrary constants. From (6.4.10) we see that

$$y_1(s) = f(s) + \lambda K y_0(s)$$

is also a solution of (6.4.7), and therefore,

$$f(s) + \lambda K y_0(s) = y_0(s) + \sum_{i=1}^{m} b_i y_i(s). \qquad (6.4.25)$$

Thus the function $\quad \bar{y}(s) = y_0(s) + \sum_i' \frac{b_i}{\rho_i - 1} y_i(s),$

where the summation $\sum'$ is taken over all eigenfunctions for which $\rho_i \neq 1$, satisfies

$$f(s) + \lambda K \bar{y}(s) = \bar{y}(s) + \sum_i'' b_i y_i(s). \qquad (6.4.26)$$

In (6.4.26), the summation $\sum''$ is over the eigenfunctions with $\rho_i = 1$. Multiplying (6.4.26) by $\bar{z}_j(s)$ and integrating, we have in view of (6.4.20) and (6.4.21),

$$0 = \sum_i'' b_i \int \bar{z}_j(s) y_i(s) \, ds = \sum_i'' b_i \gamma_{ij},$$

where the coefficients γ_{ij} are thus defined. If the determinant $|\gamma_{ij}|$ is not zero, it follows that $b_i = 0$ $(i = 1, ..., m)$, and $y(s)$ in (6.4.26) is in fact a solution of (6.4.1). We leave to the reader the construction of a modified function which will provide a solution if the determinant $|\gamma_{ij}|$ vanishes.

To sum up, we see that (6.4.1) has a solution if and only if $f(s)$ is orthogonal to all eigenfunctions of the homogeneous transposed equation (6.4.20). Thus Fredholm's third theorem is valid for equations with a kernel, some iterate of which is continuous. Also, if the homogeneous equation, and therefore its transpose, have no eigenfunctions, the non-homogeneous equation has always a unique solution. This is the statement of the Fredholm alternative for equations with a continuous iterated kernel.

6.5. Symmetric kernels. The further development of the theory of boundary value and eigenvalue problems for self-adjoint elliptic equations leads to the study of integral equations with real, symmetric kernels $k(s, t) = k(t, s)$. We shall, as in the preceding section, suppose that some iterated kernel $k^{(N)}(s, t)$ is continuous, so that our results will apply to the integral equations as they are formulated in such problems. Of particular importance are the eigenvalues and eigenfunctions of the homogeneous equation

$$y(s) = \lambda \int k(s, t) y(t) \, dt. \qquad (6.5.1)$$

In the notation of the preceding section, (6.5.1) may be written $y(s) = \lambda K y(s)$. Note that since $k(s, t)$ is symmetric, the integral equation coincides with its transposed equation, and the statements of the Fredholm theorems are simplified accordingly for symmetric kernels.

Our first result for (6.5.1) is as follows: *The eigenfunctions corresponding to distinct eigenvalues of* (6.5.1) *are orthogonal.* This is easily proved, for if $\lambda_n \neq \lambda_m$ and $y_n(s)$, $y_m(s)$ are corresponding eigenfunctions, we have

$$\lambda_n \int y_n(s)y_m(s)\,ds = \int \lambda_n y_n(s)\lambda_m K y_m(s)\,ds$$

$$= \int \lambda_n K y_n(s)\lambda_m y_m(s)\,ds$$

$$= \lambda_m \int y_n(s)y_m(s)\,ds.$$

Thus we must have

$$\int y_n(s)y_m(s)\,ds = 0, \quad \text{if} \quad \lambda_n \neq \lambda_m, \tag{6.5.2}$$

which expresses the result stated.

We may use this orthogonality property to establish that: *The eigenvalues* λ_n *of a real symmetric kernel* $k(s,t)$ *are real.* To show this, let $\lambda = \alpha + i\beta$, $\beta \neq 0$, be an eigenvalue of the kernel $k(s,t)$, so that there exists at least one eigenfunction $y(s)$ not identically zero. But, since $k(s,t)$ is real, the conjugate complex function $\bar{y}(s)$ must be an eigenfunction for the eigenvalue $\bar{\lambda} = \alpha - i\beta$ which is different from λ. But then $y(s)$ and $\bar{y}(s)$ are orthogonal:

$$\int y(s)\bar{y}(s)\,ds = \int |y(s)|^2\,ds = 0,$$

so $y(s) \equiv 0$ which is a contradiction. Therefore only real values of λ may be eigenvalues.

Since (6.5.1) is linear and homogeneous, any constant multiple of an eigenfunction is again an eigenfunction. We may therefore normalize all eigenfunctions so that their square integrals over D are equal to unity:

$$\int y_n^2(s)\,ds = 1. \tag{6.5.3}$$

If to one eigenvalue λ_n there correspond several linearly independent eigenfunctions $y_n^{(1)}(s),\dots,y_n^{(p)}(s)$, it is possible to arrange p independent linear combinations of these which are mutually orthogonal. We denote these new functions by $y_m(s)$ $(m = n,\ n+1,\dots,\ n+p-1)$, and write

$$\lambda_n = \lambda_{n+1} = \dots = \lambda_{n+p-1},$$

for each eigenvalue in turn, in the order of increasing absolute values. We finally arrive at a sequence $\lambda_1,\dots,\lambda_n,\dots$ of eigenvalues, which may be repeated, and a corresponding sequence $y_1(s),\dots,y_n(s),\dots$ of eigenfunctions, each of which is normalized according to (6.5.3), and which satisfy the orthogonality conditions (6.5.2) for every unequal pair of integers m and n. These eigenfunctions are said to form an orthonormal set.

We will next show that *a symmetric kernel* $k(s,t)$ *always has at least one eigenvalue.* By the results of the preceding section, it will be enough to show that the continuous iterated kernel $k^{(N)}(s,t)$ has at least one eigenvalue, say λ^N, for then one of the Nth roots $(\lambda^N)^{1/N}$, which must also be real, is an eigenvalue of $k(s,t)$. The existence of an eigenvalue for $k^{(N)}(s,t)$ will be shown by demonstrating that the resolvent $R_N(s,t,\lambda)$ of $k^{(N)}(s,t)$ has a pole in the finite λ-plane.

Let us expand $R_N(s,t,\lambda)$ in a power series as a function of λ. From the resolvent relations (6.1.10) or (6.1.11) we find after substituting a series with undetermined coefficients, and then equating like powers of λ, that the series has the form

$$R_N(s,t,\lambda) = \sum k^{N(n+1)}(s,t)\lambda^n. \qquad (6.5.4)$$

Here the coefficients $k^{N(n+1)}(s,t)$ are the iterates of the kernel $k^{(N)}(s,t)$ itself. We know that the poles of the resolvent are the zeros of $D_N(\lambda)$ and these zeros are independent of s and t. From (6.5.4),

$$\int R_N(s,s,\lambda)\,ds = \int \sum k^{N(n+1)}(s,s)\lambda^n\,ds$$

$$= \sum \left(\int k^{N(n+1)}(s,s)\,ds \right)\lambda^n \qquad (6.5.5)$$

$$= \sum a_n \lambda^n,$$

say, where the coefficients a_n are thus defined. Now we have, for all real values of parameters α, β,

$$0 \leqslant \iint [\alpha k^{Nn}(s,t)+\beta k^{N(n+1)}(s,t)]^2\,ds\,dt$$

$$= \alpha^2 \iint [k^{Nn}(s,t)]^2\,ds\,dt+2\alpha\beta \iint k^{Nn}(s,t)k^{N(n+1)}(s,t)\,ds\,dt+$$

$$+\beta^2 \iint [k^{N(n+1)}(s,t)]^2\,ds\,dt.$$

Since $k(s,t)$ is symmetric, we find

$$0 \leqslant \alpha^2 \int k^{2Nn}(s,s)\,ds+2\alpha\beta \int k^{2Nn+N}(s,s)\,ds+\beta^2 \int k^{2N(n+1)}(s,s)\,ds$$

$$= \alpha^2 a_{2n-2}+2\alpha\beta a_{2n-1}+\beta^2 a_{2n}.$$

This quadratic form being non-negative, we have

$$a_{2n-1}^2 \leqslant a_{2n-2}a_{2n},$$

or

$$\frac{a_{2n}}{a_{2n-1}} \geqslant \frac{a_{2n-1}}{a_{2n-2}}.$$

Thus the ratio of absolute values of terms of the series in (6.5.5) is

$$\frac{a_{2n}}{a_{2n-1}}|\lambda| \geqslant \frac{a_{2n-1}}{a_{2n-2}}|\lambda| \geqslant \cdots \geqslant \frac{a_2}{a_1}|\lambda|,$$

which exceeds unity if $|\lambda|$ is sufficiently large. The series (6.5.5) has, therefore, a finite radius of convergence, and so the resolvent $R_N(s, t, \lambda)$ has a singular point at a finite distance from the origin in the λ-plane. Thus $D_N(\lambda)$ has there a zero, and so an eigenfunction of $k^{(N)}(s, t)$ exists. Hence finally the kernel $k(s, t)$ has at least one eigenvalue λ_1 and a corresponding eigenfunction $y_1(s)$.

The result just proved makes it possible to find an expression for the symmetric kernel $k(s, t)$ in terms of its eigenvalues and eigenfunctions. We may suppose that $y_1(s)$ is normalized by (6.5.3). Then the function

$$k_1(s, t) = k(s, t) - \frac{y_1(s)y_1(t)}{\lambda_1}$$

is again a real, symmetric kernel. If $k_1(s, t)$ is not identically zero, it has an eigenvalue λ_2 and eigenfunction $y_2(s)$. Now

$$\int y_1(s)y_2(s)\, ds = \lambda_2 \int\!\!\int y_1(s)k_1(s, t)y_2(t)\, dtds$$

$$= \lambda_2 \int y_2(t)\, dt \int \left[y_1(s)k(s, t) - \frac{y_1^2(s)y_1(t)}{\lambda_1}\right] ds$$

$$= \lambda_2 \int y_2(t)\, dt\left[\frac{y_1(t)}{\lambda_1} - \frac{y_1(t)}{\lambda_1}\right] = 0.$$

Thus $y_1(s)$ and $y_2(s)$ are orthogonal, and so must be distinct. Furthermore,

$$y_2(s) = \lambda_2 \int k_1(s, t)y_2(t)\, dt$$

$$= \lambda_2 \int k(s, t)y_2(t)\, dt - \frac{\lambda_2 y_1(s)}{\lambda_1} \int y_1(t)y_2(t)\, dt$$

$$= \lambda_2 \int k(s, t)y_2(t)\, dt.$$

Hence also λ_2 and $y_2(s)$ are an eigenvalue and eigenfunction of the original kernel $k(s, t)$.

This process may be repeated, and we obtain, for each integer n,

$$k(s, t) = \sum_{i=1}^{n} \frac{y_i(s)y_i(t)}{\lambda_i} + k_n(s, t), \qquad (6.5.6)$$

where $k_n(s, t)$ is a 'remainder' after n steps. It may happen after a finite number of steps that the remainder vanishes, in which case we have expressed $k(s, t)$ as a finite bilinear sum. However, if there exist infinitely many eigenvalues and eigenfunctions the process does not terminate, and we are led to the series expansion

$$k(s,t) = \sum_{n=0}^{\infty} \frac{y_n(s)y_n(t)}{\lambda_n},$$ (6.5.7)

in which the summation extends over all eigenvalues λ_n ranged in order of increasing absolute values. The argument just given does not prove that this bilinear series is convergent; but if the convergence is assumed, the sum of the series must be the function $k(s,t)$.

6.6. Eigenfunction expansions. The eigenfunctions of a symmetric kernel $k(s,t)$ form an orthonormal set. Relative to such an orthonormal set, a function $f(s)$ defined on D has Fourier coefficients

$$f_n = \int f(s)y_n(s)\,ds,$$ (6.6.1)

and a Fourier series $\sum_{n=0}^{\infty} f_n\, y_n(s),$ (6.6.2)

which 'represents' the function $f(s)$. The Fourier expansion (6.6.2) can often be shown to be convergent or summable to the sum $f(s)$; however, we shall not enter here into the details of the convergence theory. The interest of such expansions for differential equations is that a Fourier expansion of a function given as data in an initial value problem leads to a resolution of the problem into its 'normal modes', and thence to a solution of the problem by the superposition of these modes. According to our development of the theory, these latter appear as eigenfunctions of integral equations, and it is from this point of view that we study them here. In Chapter VIII we will apply the results to initial value problems for 'separable' differential equations of parabolic or hyperbolic type.

The bilinear series (6.5.7) is in a sense a Fourier series. For, considering the kernel $k(s,t)$ as a function of s, we see that its nth Fourier coefficient is

$$\int y_n(s)k(s,t)\,ds = \frac{y_n(t)}{\lambda_n},$$

and therefore its Fourier expansion must be

$$k(s,t) = \sum_{n=1}^{\infty} \frac{y_n(s)y_n(t)}{\lambda_n}.$$ (6.6.3)

We may regard this formal series as the expression of the infinite and continuous 'matrix' $k(s,t)$ as a diagonal sum of normal modes obtained by a principal axis transformation.

The form of the Fourier series for the iterated kernels $k^{(p)}(s, t)$ may be formed from (6.6.2) by an easy formal calculation in which the orthonormal property of the functions $y_n(s)$ is utilized. The result, which the reader can easily verify, is

$$k^{(p)}(s, t) = \sum_{n=1}^{\infty} \frac{y_n(s)y_n(t)}{\lambda_n^p}.$$ (6.6.4)

For $p \geqslant N$ the iterated kernels are now assumed to be continuous.

We shall make use of the following property of the Fourier coefficients f_n of (6.6.1), known as Bessel's inequality:

$$\sum_{n=1}^{\infty} f_n^2 \leqslant \int [f(s)]^2 \, ds.$$ (6.6.5)

To establish this relation, we may assume that the integral on the right converges. Let us expand out the non-negative expression

$$\int \left[f(s) - \sum_{n=1}^{k} f_n y_n(s) \right]^2 ds$$

$$= \int [f(s)]^2 \, ds - 2 \sum_{n=1}^{k} f_n \int f(s)y_n(s) \, ds + \sum_{n=1}^{k} f_n^2$$

$$= \int [f(s)]^2 \, ds - \sum_{n=1}^{k} f_n^2 \geqslant 0.$$

Here we have used (6.6.1) and the orthonormal property of the $y_n(s)$. Thus the sum of squares of the Fourier coefficients is bounded above by an integral which is independent of k. Letting k tend to infinity, it follows that the series converges, and that (6.6.5) holds.

Now the Fourier coefficients of $k^{(N)}(s, t)$ are $y_n(t)/\lambda_n^N$; applying the Bessel inequality we see that

$$\sum_{n=1}^{\infty} \frac{y_n^2(t)}{\lambda_n^{2N}} \leqslant \int [k^{(N)}(s, t)]^2 \, ds.$$ (6.6.6)

Therefore, integrating with respect to t and using (6.5.3), we find

$$\sum_{n=1}^{\infty} \frac{1}{\lambda_n^{2N}} \leqslant \int \int [k^{(N)}(s, t)]^2 \, ds.$$ (6.6.7)

In this series, each eigenvalue λ_n appears a number of times equal to its multiplicity as an eigenvalue. The formula shows that every eigenvalue has a finite multiplicity, and that the absolute values of the λ_n tend to infinity with n.

As a second application of the Bessel inequality, we will show that a certain class of functions $f(s)$ have convergent Fourier expansions (6.6.2).

These are functions of the form

$$f(s) = \int k^{(N)}(s, t)g(t) \, dt, \tag{6.6.8}$$

where $g(t)$ is of integrable square over D (or, in particular, is continuous). Such functions are said to be representable by the kernel $k^{(N)}(s, t)$. Now formally, we have

$$f(s) = \int \sum_{n=1}^{\infty} \frac{y_n(s)y_n(t)}{\lambda_n^N} g(t) \, dt$$

$$= \sum_{n=1}^{\infty} \frac{y_n(s)}{\lambda_n^N} \int y_n(t)g(t) \, dt \tag{6.6.9}$$

$$= \sum_{n=1}^{\infty} \frac{y_n(s)}{\lambda_n^N} g_n.$$

This Fourier series is convergent at every point s, since

$$\left| \sum_{n=m}^{M} \frac{y_n(s)}{\lambda_n^N} g_n \right| \leqslant \left[\sum_{n=m}^{M} g_n \right]^{\frac{1}{2}} \left[\sum_{n=m}^{M} \frac{y_n(s)}{\lambda_n^{2N}} \right]^{\frac{1}{2}}.$$

The first term on the right tends to zero as m, M tend to infinity, by Bessel's inequality applied to $g(s)$. As for the second factor, it is bounded independently of m and M in view of (6.6.6). Since the remainder term on the left tends to zero as m, M tend to infinity, the convergence of (6.6.9) follows. In the applications of this result in Chapter VIII, it will be seen that a function $f(s)$ which possesses a sufficient number of derivatives can be represented by the appropriate kernel as in (6.6.8).

Exercise 1. Is (6.6.9) uniformly convergent with respect to s?

Exercise 2. Prove directly that the sum of the series in (6.6.9) is actually $f(s)$.

VII

BOUNDARY VALUE PROBLEMS

WE may now apply the theory developed in Chapter VI to the solution of those problems concerning the self-adjoint elliptic equation

$$\Delta u = q(P)u,$$

which were formulated in Chapter V by means of integral equations. It is necessary here to assume that the coefficient $q(P)$ is non-negative, and treatment of the contrary case is postponed to the next chapter. The first step is the solution of the Poisson equation in a closed space of the type considered in § 5.5. From this result it follows that a symmetric fundamental solution in the large may be constructed. In the second part of this chapter the actual existence proofs for the boundary value problems of Dirichlet, Neumann, and Robin are presented. Here the fundamental solution is used to construct single and double layer potentials.

The remainder of the chapter concerns the definition and properties of certain special fundamental solutions which are related to the various boundary value problems on a given domain. These Green's functions may be used to construct the solutions of the boundary value problems, and also play a considerable role in the following chapter. By means of such functions we also define the reproducing kernel function, and deduce from it a convergence theorem for solutions of the elliptic differential equation.

7.1. Poisson's equation and the fundamental solution in the large.

In § 5.4 we saw that the solution of Poisson's equation

$$Lu = \Delta u - qu = f(P) \tag{7.1.1}$$

could be expressed as a problem in integral equations, and we were able to solve the integral equation by successive approximations provided that the domain was small enough. Let us now investigate this same problem on a closed Riemannian space F, and apply Fredholm's theory to solve the integral equation. We shall require a uniqueness theorem such as that proved in § 5.5 when $q \geqslant 0$, $q \not\equiv 0$, and we therefore assume that the coefficient $q(P)$ satisfies these restrictions.

Let the closed space be denoted by F. We have the parametrix

$$\omega(P, Q) = \frac{\rho(s/\epsilon)s^{-N+2}}{\omega_N(N-2)}, \tag{7.1.2}$$

where the function $\rho(x)$ is included in order to smooth off the values of $\omega(P, Q)$ to zero when $s(P, Q)$ exceeds a certain positive limit ϵ. Thus, we choose

$$\rho(x) = \begin{cases} 1 & 0 \leqslant x \leqslant \tfrac{1}{2} \\ 0 & x \geqslant 1 \end{cases}$$

and require that $\rho(x)$ should be sufficiently often differentiable for all values of its argument. As in § 5.3 we define the integral operator

$$\Omega u(P) = \int_F \omega(P, Q)u(Q)\, dV_Q, \tag{7.1.3}$$

the domain of integration being now the whole space F. Similarly, writing

$$q(P, Q) = -L_P\, \omega(P, Q), \tag{7.1.4}$$

we define

$$\Phi u(P) = \int_F q(P, Q)u(Q)\, dV_Q. \tag{7.1.5}$$

The transpose Φ' of the integral operator Φ is

$$\Phi' u(P) = \int_F q(Q, P)u(Q)\, dV_Q. \tag{7.1.6}$$

As in § 5.3 the operators Ω and Φ satisfy the relation

$$L\Omega u(P) = u(P) - \Phi u(P); \tag{7.1.7}$$

indeed, the proof is quite unaltered.

Following our earlier procedure, let us take as a trial solution of the Poisson equation

$$u(P) = \Omega v(P), \tag{7.1.8}$$

then, as before, we see that $v(P)$ must satisfy the integral equation

$$v(P) - \Phi v(P) = f(P). \tag{7.1.9}$$

By Fredholm's third theorem (7.1.9) has a solution if and only if $f(P)$ is orthogonal to all solutions $z(P)$ of the homogeneous transposed equation, namely,

$$z(P) - \Phi' z(P) = 0. \tag{7.1.10}$$

The function $q(P, Q)$ which appears as kernel of these integral equations has a singularity of order $N-2$ when $P \to Q$. Thus, from § 6.4, the iterated kernels have singularities of orders $N-4$, $N-6$; and the iterated kernels of orders exceeding $\tfrac{1}{2}N$ are continuous. Therefore (7.1.10) has at most a finite number k of linearly independent eigenfunctions $z_\nu(P)$ ($\nu = 1, ..., k$). If we form the k functions $Lz_\nu(P)$, they also must be linearly independent, since, if they were not, there would exist a non-zero solution of the equation $Lz = \Delta z - qz = 0$ in F; and this is impossible because $q \geqslant 0$, $q \not\equiv 0$. Likewise, the k functions $L^2 z_\nu(P)$ are linearly independent.

Exercise. Show that the matrix $(\alpha_{\mu\nu})$, where

$$\alpha_{\mu\nu} = \int_F Lz_\mu \, Lz_\nu \, dV = \int_F L^2 z_\mu . z_\nu \, dV, \qquad (7.1.11)$$

is positive definite and therefore non-singular.

If the homogeneous transposed equation (7.1.10) has solutions, we cannot solve the integral equation (7.1.9) unless the corresponding conditions of orthogonality are satisfied. Let us therefore modify the trial solution so that these conditions can be met. We shall take as a new trial solution

$$u(P) = \Omega v(P) + \sum b_\nu \, Lz_\nu(P), \qquad (7.1.12)$$

where the $z_\nu(P)$ are the eigenfunctions themselves, and the k coefficients b_ν are as yet undetermined. Now,

$$\begin{aligned} Lu(P) &= L\Omega v(P) + \sum b_\nu \, L^2 z_\nu(P) \\ &= v(P) - \Phi v(P) + \sum b_\nu \, L^2 z_\nu(P), \qquad (7.1.13) \end{aligned}$$

so, if the Poisson equation is to hold, we must satisfy the modified integral equation

$$v(P) - \Phi v(P) = f(P) - \sum b_\nu \, L^2 z_\nu(P). \qquad (7.1.14)$$

The kernel of this equation is again the function $q(P, Q)$. Let us choose the k numbers b_ν to satisfy the necessary orthogonality conditions, which take the form

$$\int \left[f(P) - \sum_\nu b_\nu \, L^2 z_\nu(P) \right] z_\mu(P) \, dV_P = 0 \quad (\mu = 1, \ldots, k). \qquad (7.1.15)$$

That is, let

$$\begin{aligned} a_\mu &= \int f(P) z_\mu(P) \, dV_P = \int \sum_\nu b_\nu \, L^2 z_\nu(P) z_\mu(P) \, dV_P \\ &= \sum_\nu b_\nu \int L^2 z_\nu(P) z_\mu(P) \, dV_P = \sum_\nu b_\nu \, \alpha_{\mu\nu}. \end{aligned} \qquad (7.1.16)$$

Thus (7.1.16) constitutes a set of k linear algebraic equations to be solved for the b_ν. In view of the preceding exercise, the matrix of the coefficients is non-singular, and the b_ν are therefore well-determined by (7.1.16). Then the orthogonality conditions relative to the modified equation (7.1.14) are satisfied. Hence, finally, $u(P)$ defined by (7.1.12) is a solution of the Poisson equation $Lu(P) = f(P)$.

That this solution must be unique follows from the condition $q \geqslant 0$, $q \not\equiv 0$; for the difference of any two solutions of the Poisson equation is a solution in F of the homogeneous differential equation, and so must be identically zero. We have therefore proved the following theorem: *The Poisson equation*

$$\Delta u(P) - q(P)u(P) = f(P) \tag{7.1.1}$$

in a closed space F, $q(P) \geqslant 0$, $q(P) \not\equiv 0$, has a unique solution.

Turning now to examine the form of this solution, we see that the terms of the solution function (7.1.12) involve the solution $v(P)$ of the integral equation and also the constants b_ν ($\nu = 1,...,k$). It is a straightforward matter to verify that all of these are expressed as integrals over the space F, with the function $f(P)$ appearing linearly in the integrand. Thus

$$u(P) = - \int_F \gamma(P, Q)f(Q) \, dV_Q, \tag{7.1.17}$$

where the kernel $\gamma(P, Q)$ can be expressed in terms of the parametrix $\omega(P, Q)$, the function $q(P, Q)$, and the eigenfunctions $z_\nu(P)$. While we need not find the exact expression for $\gamma(P, Q)$, we see from the term $\Omega v(P)$ in (7.1.12) that $\gamma(P, Q)$ has the same singularity as the parametrix. Furthermore, the argument used in § 5.3 shows here as well that if (7.1.17) holds for every function $f(P)$, then

$$L_P \gamma(P, Q) = 0 \quad (P \neq Q). \tag{7.1.18}$$

Finally, we shall establish that $L_Q \gamma(P, Q) = 0$, by showing that $\gamma(P, Q)$ is symmetric in its two argument points:

$$\gamma(\dot{P}, Q) = \gamma(Q, P). \tag{7.1.19}$$

To prove this, let $K_\epsilon(P)$, $K_\epsilon(Q)$ be geodesic spheres of radius ϵ about P and Q, and consider the integral

$$I(\epsilon) = \int_{K_\epsilon(P)+K_\epsilon(Q)} \left[\gamma(r, P)\frac{\partial \gamma(r, Q)}{\partial n_r} - \gamma(r, Q)\frac{\partial \gamma(r, P)}{\partial n_r} \right] dS_r.$$

By Green's formula, $I(\epsilon)$ can be expressed as a volume integral extended over F, less the interiors of the two spheres. Since (7.1.18) holds, it follows that this integral vanishes. Thus $I(\epsilon) = 0$. On the other hand, we may evaluate the surface integrals in the limit as $\epsilon \to 0$, making use of the asymptotic behaviour of $\gamma(P, Q)$ as the two argument points coalesce. The calculation has already been done in (5.3.10); let us state the result here. From the surface $K_\epsilon(P)$ the first term yields zero, the second term yields $-\gamma(P, Q)$. The surface $K_\epsilon(Q)$ gives us from the first term $\gamma(P, Q)$, from the second term zero. Since $I(\epsilon)$ is zero, we must conclude that $\gamma(P, Q) = \gamma(Q, P)$, which is just (7.1.19).

To sum up, then, $\gamma(P, Q)$ is a symmetric fundamental solution in the large for the differential equation in F. We may call $\gamma(P, Q)$ the Green's function of the closed space F.

Exercise. Prove that the Green's function of F is unique, and is non-negative.

7.2. Solution of the boundary value problems.

Let D be a region, with boundary B, of a Riemannian space, and let us consider the Dirichlet and Neumann problems for the equation

$$\Delta u(P) = q(P)u(P) \quad (q(P) > 0), \tag{7.2.1}$$

on D. First, let F be the double of the given domain D. We may suppose that the differential equation is defined on the whole of F, and that $q(P)$ is positive in the complementary domain $\bar{D}$. Let $\gamma(P, Q)$ be the Green's function of F. Thus $\gamma(P, Q)$ is a fundamental solution in the large defined in D and also in the exterior region $\bar{D}$ of F.

From § 5.6 we recall that if

$$k(p, q) = 2\frac{\partial \gamma(p, q)}{\partial n_q}, \tag{7.2.2}$$

then the Dirichlet problem for D may be reduced to the integral equation

$$f(p) = \varphi(p) - \lambda \int_B k(p, q)\varphi(q) \, dS_q, \tag{7.2.3}$$

where $f(p)$ represents the assigned boundary values for the solution. In (7.2.3), we have $\lambda = +1$ for the interior problem, $\lambda = -1$ for the exterior problem. Similarly, the Neumann problem was formulated by means of the transposed integral equation

$$g(p) = \psi(p) - \lambda \int_B k(q, p)\psi(q) \, dS_q, \tag{7.2.4}$$

where $\lambda = +1$ for the exterior Neumann problem and $\lambda = -1$ for the interior problem. Thus, for each of these four boundary value problems, the transposed equation is that for the 'opposite' problem for the complementary domain. In particular, the homogeneous transposed equation arises in connexion with the opposite problem, with zero data, for the complementary domain.

From (5.6.18) we see that the singularity of the kernel $k(p, q)$ is of order $N-2$. Since the dimension of the closed domain of integration B is $N-1$, it follows that the successive iterated kernels are singular of orders $N-2$, $N-3,...$, and that the Nth iterated kernel is continuous. We may therefore apply the Fredholm theorems to (7.2.3) and (7.2.4). In order to avoid repetition, we shall assume, once for all, that the data assigned are continuous.

We shall demonstrate that the values $\lambda = +1$ and $\lambda = -1$ are not eigenvalues of the kernel $k(p,q)$. Then, by the Fredholm alternative, it will follow that the integral equations (7.2.3) and (7.2.4) possess solutions for every assigned set of boundary values.

Let us suppose, then, that $\lambda = +1$ is an eigenvalue; that is, that there exists a solution $\psi(p)$, not identically zero, of the homogeneous equation

$$0 = \psi(p) - \int_B k(q, p)\psi(q) \, dS_q. \qquad (7.2.5)$$

Corresponding to this eigenfunction $\psi(p)$, there exists a single layer potential

$$V(P) = \int_B \psi(q)\gamma(P, q) \, dS_q. \qquad (7.2.6)$$

From § 5.6 we see that $V(P)$ has, as limiting value from the exterior region, the normal derivative zero:

$$\frac{\partial V(p)}{\partial n_+} = 0. \qquad (7.2.7)$$

But $V(P)$ is a solution of the homogeneous differential equation in $\tilde{D}$. Then, since the Neumann problem for $\tilde{D}$ has at most one solution, it must follow that $V(P) \equiv 0$ in $\tilde{D}$. In particular, $V(P)$ is zero on the boundary B. and, being a single layer, is continuous across B. Hence $V(p)$ is also zero, Again, in the interior of D, the single layer potential $V(P)$ is a solution of the homogeneous differential equation. In view of the uniqueness of a solution of the Dirichlet problem in D, it follows that $V(P) \equiv 0$ in D. Consequently,

$$\frac{\partial V(p)}{\partial n_-} = 0, \qquad (7.2.8)$$

and from (7.2.7) we see that the discontinuity across B of the normal derivative of $V(P)$ is zero. However, this implies, by (5.4.13), that the single layer density $\psi(p)$ is itself zero. Therefore no non-zero eigenfunction corresponding to the value $\lambda = +1$ can exist.

Since $\lambda = +1$ is not an eigenvalue, the non-homogeneous equations

(7.2.3) and (7.2.4) have unique solutions. Retracing the formulation of the boundary value problems in § 5.6, we conclude that in a closed space F the interior Dirichlet and exterior Neumann problems have unique solutions.

Next let us consider the case $\lambda = -1$. An argument entirely similar to that just given carries us through this case as well. However, we may reduce the exterior Dirichlet and interior Neumann problems to the interior Dirichlet and exterior Neumann problems, simply by relabelling the two domains D and $\check{D}$ which together constitute the closed space F. When this is done, the interior and exterior domains are interchanged, and the proof above applies word for word.

Indeed the argument given above will be valid provided that D is any region of some closed space F, not necessarily the double of D. In fact, D need not be connected but may consist of several distinct regions. We therefore formulate the general boundary value theorem: *Let D be a region with a smooth boundary surface B, interior to a closed space F. Then the Dirichlet and Neumann problems for (7.2.1), both for D and the complementary region exterior to D in F, have unique solutions.*

The properties of double layer potentials make possible a similar proof in which the existence of an eigenfunction is shown to lead to a contradiction. We suggest the construction of this proof as an exercise.

The distinction between the interior and exterior boundary value problems becomes essential, however, if the Riemannian space upon which our differential equation is defined is an infinite, open space V_N, such as Euclidean space, or a space of constant negative curvature. In this case, some restriction on the behaviour at infinity of the solution functions is necessary to ensure uniqueness of the solutions for the exterior problems. A condition sufficient for this purpose is that the solution $u(P)$ should tend to zero as $P \to \infty$, rapidly enough that

$$\int_{B_n} u \frac{\partial u}{\partial n} \, dS \to 0 \quad (n \to \infty), \qquad (7.2.9)$$

for some sequence of surfaces B_n, enclosing D, which tend to infinity. For instance, it is sufficient, in Euclidean space, if $|u(P)|+|\nabla u(P)| < Kr^{-\frac{1}{2}N}$ as $r \to \infty$. To verify this assertion, suppose that $u(P)$ is a solution function satisfying the condition (7.2.9), and that, for instance, $u(p) = 0$ or $\partial u(p)/\partial n = 0$ on B, so that $\int_B u(\partial u/\partial n) \, dS$ is zero. Let D_n denote the region enclosed by B_n and B, and let us evaluate the Dirichlet integral of $u(P)$

over D_n. From Green's formula we see that

$$E_n(u) = \int\limits_{B-B_n} u\,\frac{\partial u}{\partial n}\,dS \to 0 \quad (n \to \infty),$$

and therefore that $E(u)$, the Dirichlet integral over the whole space, is zero. Thus also must $u(P)$ vanish identically.

We shall also assume that a fundamental solution $\gamma(P, Q)$ is defined in the whole space, and that as a function of P, $\gamma(P, Q)$ itself satisfies (7.2.9), uniformly with respect to Q, for Q in any bounded domain. This supposition is fulfilled for the equation

$$\Delta u = k^2 u,$$

in Euclidean space of N dimensions. Indeed, from § 4.4, we see that this equation has a fundamental solution which will be a Bessel function with pure imaginary argument. This solution may be expressed by means of the modified Bessel function of the second kind $K_n(x)$, in the form

$$\gamma(P, Q) = \frac{1}{r^{\frac{1}{2}(N-2)}} K_{\frac{1}{2}(N-2)}(kr) \qquad (r = r(P, Q)).$$

As $r \to \infty$, we have the asymptotic formula

$$K_n(kr) \sim \sqrt{\left(\frac{\pi}{kr}\right)}\, e^{-kr},$$

so that $\gamma(P, Q)$ is exponentially small at great distances. For $N = 2$, the solution is $K_0(kr)$, which has a logarithmic singularity at the origin, while for $N = 3$,

$$\gamma(P, Q) = \frac{e^{-kr}}{4\pi r}.$$

Let us consider the interior Dirichlet and exterior Neumann problems for the finite domain D in the infinite open space V_N. Again we may formulate the problems by means of the integral equations (7.2.3) and (7.2.4); this part of the work proceeds just as before. If, then, we can demonstrate that $\lambda = +1$ is not an eigenvalue, we may again conclude that the boundary value problems in question are uniquely solvable.

If $\lambda = +1$ is an eigenvalue, there exists a single layer potential $V(P)$, defined again by (7.2.6), whose normal derivative (7.2.7) vanishes according to (7.2.5). Our task is now to show that $V(P)$ is identically zero in the exterior region. Since $V(P)$ is a solution function in this region, which has a vanishing normal derivative on B, it is sufficient to show that $V(P)$ satisfies the order of magnitude condition (7.2.9). But we have assumed that the fundamental solution $\gamma(P, q)$ satisfies this condition (7.2.9), uniformly with respect to q, as q ranges over B. Thus

$$|V(P)| = \left| \int_B \gamma(P, q)\psi(q) \, dS_q \right|$$

$$\leqslant \max_B |\gamma(P, q)| \int |\psi(q)| \, dS_q$$

$$\leqslant K\gamma(P),$$

say, where $\gamma(P)$ is a function of P which also satisfies (7.2.9). Therefore $V(P)$ satisfies (7.2.9), and so $V(P) \equiv 0$ in the exterior region.

As in our earlier proof we can now show that $V(P)$ is zero in D, and that $\psi(q)$ is zero. Hence $\lambda = +1$ is not an eigenvalue. A similar argument shows that $\lambda = -1$ is not an eigenvalue. Under the hypotheses we have made, therefore, the interior and exterior Dirichlet and Neumann problems for a bounded domain D possess unique solutions satisfying (7.2.9).

The Robin problem, in which the mixed boundary condition

$$\frac{\partial u(p)}{\partial n} + h(p)u(p) = f(p) \quad (h(p) \geqslant 0, \, h(p) \not\equiv 0), \tag{7.2.10}$$

is to be satisfied, may be treated even more directly than the two preceding cases. Let us consider the interior problem, which arises in connexion with equilibrium problems of heat conduction. We try to satisfy (7.2.10) by taking as trial solution a single layer potential

$$V(P) = \int_B \sigma(q)\gamma(P, q) \, dV_q. \tag{7.2.11}$$

From (5.4.13) we find that

$$\frac{\partial V(p)}{\partial n_-} + h(p)V(p) = \tfrac{1}{2}\sigma(p) + \int_B \sigma(q)\left[\frac{\partial \gamma(p, q)}{\partial n} + h(p)\gamma(p, q)\right] dS_q, \tag{7.2.12}$$

and it follows that the boundary condition (7.2.10) leads to the integral equation

$$f(p) = \tfrac{1}{2}\sigma(p) + \int_B \sigma(q)\left[\frac{\partial \gamma(p, q)}{\partial n} + h(p)\gamma(p, q)\right] dS_q. \tag{7.2.13}$$

Equation (7.2.13) is to be solved for the undetermined single layer density $\sigma(q)$.

By the Fredholm alternative (7.2.13) has a solution provided that the corresponding homogeneous equation has no solution not identically zero. Clearly a solution of (7.2.13) in which $f(p)$ has been replaced by zero is a function $\sigma(q)$ which, inserted in (7.2.11), leads to a single layer potential $V(P)$ satisfying the homogeneous boundary condition

$$\frac{\partial V(p)}{\partial n_-} + h(p)V(p) = 0. \tag{7.2.14}$$

Since $V(P)$ is a solution of the differential equation in D, equation (5.1.12) shows that $V(P)$ is identically zero in D. Now $V(P)$ is continuous across B, and so has the boundary values zero from the exterior region. (If the exterior region is unbounded, we shall assume that (7.2.9) holds for $V(P)$.) Again, therefore, $V(P)$ must be identically zero in the exterior region. Consequently, by the reasoning based on (5.4.13) which we have used before, the density $\sigma(p)$ of the single layer potential $V(P)$ must be identically zero. That is to say, the homogeneous integral equation corresponding to (7.2.13) has no solution other than zero. By Fredholm's alternative, we may therefore conclude that (7.2.13) itself has a unique solution $\sigma(p)$. Finally, it is seen that the single layer potential (7.2.11), constructed with this solution $\sigma(p)$ as single layer density, is a solution of the Robin problem for the domain D. We conclude that the Robin problem (7.2.10) for D has a unique solution.

Exercise 1. Investigate the Robin problem for the exterior region in an infinite space.

Exercise 2. Show that, in an infinite space, a solution of the exterior Dirichlet problem is unique, provided that the solution tends to zero at infinity. *Hint*: Use the maximum principle.

Exercise 3. Do the proofs given here apply to Laplace's equation? Which of the problems considered fail to have unique solutions in the case of harmonic functions?

7.3. Representation formulae. The existence theorems of the preceding section do, in principle, furnish a construction of the solution functions. However, the formulae obtained by carrying out the various steps of these proofs would be prohibitively complicated, and a more direct construction of the solutions is possible. We shall devote the remainder of this chapter to this simpler method, which in turn leads to further developments of a theoretical nature. Only the interior boundary value problems will be considered.

The existence proofs show us that the solutions of the various boundary value problems are given by single or double layer potentials, the layer densities of which are determined by the solution of integral equations. Our method will consist of making a particular choice of the fundamental solution for each type of problem, so that the layer densities which we require in these potentials are none other than the given boundary value data of the problem. We start with the formula (5.4.6), namely,

$$u(P) = \int\limits_B \left\{ \gamma(P,q) \frac{\partial u(q)}{\partial n_q} - u(q) \frac{\partial \gamma(P,q)}{\partial n_q} \right\} dS_q. \qquad (7.3.1)$$

We recall that in this formula $\gamma(P,Q)$ is any fundamental solution in the large. This formula represents any solution $u(P)$ of the differential equation as the sum of a single layer and of a double layer, the densities being $\partial u(p)/\partial n$ and $-u(p)$ respectively.

Consider first the Dirichlet problem for the region D. Let us write

$$G(P,Q) = \gamma(P,Q) + g(P,Q), \qquad (7.3.2)$$

where $\gamma(P,Q)$ is a fixed fundamental solution, and let us try to choose $g(P,Q)$ so that $G(P,Q)$ vanishes for Q on the boundary; and also so that $G(P,Q)$ itself will be a fundamental solution. These two requirements can be met if we choose the as yet undetermined function $g(P,Q)$ in the following way: Let $g(P,Q)$, as a function of Q, be the solution of the Dirichlet problem for D which has the boundary values $-\gamma(P,q)$. Thus $g(P,Q)$ is a solution of the differential equation as a function of Q, and depends on P through its boundary values. Now we shall need to use the reciprocal theorem (5.1.13), which for the two solution functions $u(P)$ and $g(P,Q)$ reads

$$0 = \int\limits_B \left\{ g(P,q) \frac{\partial u(q)}{\partial n_q} - u(q) \frac{\partial g(P,q)}{\partial n_q} \right\} dS_q. \qquad (7.3.3)$$

Let us add (7.3.1) and (7.3.3), and note that the function $G(P,Q)$, as defined by (7.3.2), vanishes when Q lies on the boundary B. We obtain in this way the representation formula

$$u(P) = - \int\limits_B u(q) \frac{\partial G(P,q)}{\partial n_q} dS_q, \qquad (7.3.4)$$

for the solution of the Dirichlet problem with assigned boundary values $u(p) = f(p)$.

The auxiliary function $G(P,Q)$ is known as the Green's function for the domain D and the differential equation $Lu = 0$. It is a domain functional: a function of two variables which depends not only on the differential equation but on the particular domain as well. Actually, the Green's function is uniquely determined by the differential equation and the domain. This might be verified by examining the various steps in the definition of $G(P, Q)$, but it is also easy to show directly. Indeed, if $G_1(P, Q)$ and $G_2(P, Q)$ are distinct Green's functions for the given domain, then $G_1(P, Q) - G_2(P, Q)$ is a regular and bounded solution of the differential equation, as function of Q, which vanishes on the boundary, and so must vanish identically.

The Green's function is also non-negative:

$$G(P, Q) \geqslant 0. \tag{7.3.5}$$

To show this, let P be a fixed point; then $G(P, Q)$, being positively infinite for $P = Q$, is positive in a region containing P. Also, $G(P, Q)$ is zero for Q on the boundary. If $G(P, Q)$ were negative in any region, $G(P, Q)$ would necessarily have a negative minimum in the interior of that region. This is expressly forbidden by the maximum principle, and so $G(P, Q)$ is non-negative.

Like the Green's function of a closed space, the Green's function of a bounded domain has the striking property of symmetry:

$$G(P, Q) = G(Q, P).$$

Indeed, the proof of this is analogous to that given in section 1. We consider an integral of the form

$$I(\epsilon) = \int_{B - K_\epsilon(P) - K_\epsilon(Q)} \left\{ G(P, r) \frac{\partial G(Q, r)}{\partial n_r} - G(Q, r) \frac{\partial G(P, r)}{\partial n_r} \right\} dS_r.$$

By Green's formula, $I(\epsilon)$ is equal to a volume integral over D, less the volumes of the spheres of radius ϵ about the fixed points P and Q; and this volume integral is zero. What is more, $G(P, r)$ and $G(Q, r)$ both vanish when r is a point of the boundary surface B; thus the surface integral over B also yields zero. We are left with the integrals over the surfaces of the two geodesic spheres: $S_\epsilon(P)$ and $S_\epsilon(Q)$. Let us consider the former of these two integrals. The calculation of the two terms in the integral again follows the calculations leading to (5.3.5). In fact, we see that the first term tends to zero with ϵ, and the second term contributes in the limit $- G(Q, P)$, the argument r being replaced by its limit P. A similar evaluation of the integral over $S_\epsilon(Q)$ shows that the first term contributes $G(P, Q)$ to the limit as $\epsilon \to 0$; the second term zero. Finally, therefore, the value of $I(\epsilon)$ is, in the limit, $G(P, Q) - G(Q, P)$. Since this limit is zero, the symmetry is proved. This shows, what we have not yet established, that

$G(P, Q)$ is a solution of the differential equation as a function of P, as well as a solution as function of Q.

Since the Green's function is a fundamental solution we can set $\gamma(P, Q) = G(P, Q)$ in (5.3.5). We find that, for P in D,

$$v(P) = - \int_D G(P, Q) f(Q) \, dV_Q - \int_B v(q) \frac{\partial G(P, q)}{\partial n_q} \, dS_q, \qquad (7.3.6)$$

where $f(P)$ is defined by $\quad Lv(P) = f(P).$ \qquad (7.3.7)

In these formulae, $v(P)$ is any sufficiently differentiable function, and $v(q)$ denotes its values on the boundary B. From (7.3.4) we see that the second term of (7.3.6) is a solution of the differential equation $Lu(P) = 0$ with boundary values $v(p)$. Consequently the boundary values of the first term, the volume integral, must be zero. This result, which may seem obvious, is not trivial. The reason is that, as P tends to the boundary, the Green's function does not tend to zero uniformly because of its singularity. It can be shown directly that the limit of the integral is zero, however.

Conversely, if $f(P)$ is a differentiable function, we see from (7.3.7) that the integral

$$u(P) = \int_D G(P, Q) f(Q) \, dV_Q, \qquad (7.3.8)$$

which vanishes on the boundary, is a solution of the Poisson equation $Lu(P) = -f(P)$. This solution is clearly unique. These various properties of the Green's function will be applied in our subsequent work.

Exercise. Give a physical interpretation of the Green's function as the potential of a certain point charge. What is implied, in this interpretation, by the symmetry of $G(P, Q)$?

For the Neumann problem, an entirely similar representation can be developed. In this case the domain functional is known as the Neumann function, and is denoted by $N(P, Q)$. In order to define the Neumann function we write, as in (7.3.2),

$$N(P, Q) = \gamma(P, Q) + n(P, Q), \qquad (7.3.9)$$

and choose $n(P, Q)$, as a function of Q, to be that solution of the Neumann problem for D whose normal derivative is equal to the negative of the normal derivative of $\gamma(P, q)$ on B. Thus the boundary condition for $N(P, Q)$ is

$$\frac{\partial}{\partial n_q} N(P, q) = 0. \qquad (7.3.10)$$

Let us now recall that equations (7.3.1) and (7.3.3), with $g(P, q)$ replaced by $n(P, q)$, are both valid in the present case. Adding them together, and taking account of (7.3.9) and (7.3.10), we obtain the representation formula for the Neumann problem, namely,

$$u(P) = \int\limits_B N(P, q) \frac{\partial u(q)}{\partial n_q} \, dS_q. \qquad (7.3.11)$$

Equation (7.3.11) represents the solution of the Neumann problem with assigned boundary values $\partial u(q)/\partial n = f(p)$.

The Neumann function is also uniquely determined, the verification of this being exactly the same as for the Green's function. We shall prove that $N(P, Q)$ is non-negative, by an argument rather different from that used for the Green's function. Let the point P be fixed; clearly $N(P, Q)$ is positive in some neighbourhood of P because $\gamma(P, Q) \to +\infty$ as $Q \to P$. Let us suppose that $N(P, Q)$ is negative in some non-empty part D_- of D. By what precedes, P does not lie in D_-, and so $N(P, Q)$ is a bounded regular function of Q in D_-. The boundary B_- of D_- will consist of part of the boundary of D, upon which (7.3.10) holds, and also of smooth level curves $N(P, Q) = 0$. Since D is non-empty, by hypothesis, the Dirichlet integral with respect to Q of $N(P, Q)$ over D, which we shall denote by

$$E_-(N(P, Q), N(P, Q))$$

is positive. However, we see from Green's formula that this Dirichlet integral is equal to

$$\int\limits_{B_-} N(P, q) \frac{\partial N(P, q)}{\partial n_q} \, dS_q - \int\limits_{D_-} N(P, Q) L_Q N(P, Q) \, dV_Q.$$

The integral over B_- is zero because either $N(P, q)$ or the normal derivative (7.3.10) is zero on B_-, and the integral over D_- is zero because $N(P, Q)$ is a solution of the differential equation in D_-. Therefore the assumption that $N(P, Q)$ is somewhere negative implies a contradiction.

Exercise. Use this method to show that $G(P, Q) \geqslant 0$.

Like the Green's function, the Neumann function is symmetric:

$$N(P, Q) = N(Q, P). \qquad (7.3.12)$$

The proof follows the same lines, with the Neumann function replacing the Green's function in the definition of the integral $I(\epsilon)$. The details we leave as an exercise to the reader. In view of the symmetry property, $N(P, Q)$, for $P \neq Q$, is a solution of the differential equation as a function of P.

Exercise. Using (5.3.5) and (7.3.11) show that

$$u(P) = \int_D N(P, Q) f(Q)\, dV_Q \qquad (7.3.13)$$

is the unique solution of $Lu(P) = -f(Q)$ which satisfies the boundary condition $\partial u(p)/\partial n = 0$.

Again, a similar representation holds for the Robin problem. If we denote the corresponding 'Robin function' by $R(P, Q)$, then the form of $R(P, Q)$ is

$$R(P, Q) = \gamma(P, Q) + r(P, Q), \qquad (7.3.14)$$

where $r(P, Q)$ is a solution of a certain Robin problem in the variable Q. This problem is so formulated that we have

$$\frac{\partial R(P, q)}{\partial n_q} + h(q) R(P, q) = 0. \qquad (7.3.15)$$

We leave as an exercise the details of this proof of existence, and also the verification that $R(P, Q)$ is unique.

It follows by adding (7.3.1) and the formula (7.3.3) in which $g(P, Q)$ has been replaced by $r(P, Q)$, that any solution $u(P)$ of the differential equation $Lu(P) = 0$ satisfies

$$u(P) = \int_B \left\{ R(P, q) \frac{\partial u(q)}{\partial n} - u(q) \frac{\partial R(P, q)}{\partial n} \right\} dS_q$$

$$= \int_B R(P, q) \left\{ \frac{\partial u(q)}{\partial n} + h(q) u(q) \right\} dS_q \qquad (7.3.16)$$

in view of (7.3.15). It is seen, therefore, that the solution function $u(P)$ of the Robin problem

$$\frac{\partial u(p)}{\partial n} + h(p) u(p) = f(p) \quad (h(p) \geqslant 0) \qquad (7.3.17)$$

is given by

$$u(P) = \int_B R(P, q) f(q)\, dS_q. \qquad (7.3.18)$$

This is the representation formula for the Robin problem.

A slight modification, which we shall leave to the reader, of the proof that $N(P, Q)$ is non-negative shows that $R(P, Q)$ is also non-negative. Again, the symmetry property holds: viz. $R(P, Q) = R(Q, P)$, and the proof is as before with only minor changes. Since $R(P, Q)$ depends on the non-negative function $h(p)$, we see that there exist infinitely many symmetric fundamental solutions in the large for our differential equation, namely, the Robin functions of all possible Robin problems, with non-negative functions $h(p)$.

Exercise 1. Prove that, if $f(P) \in C^2$ but is otherwise arbitrary,

(a) $\quad E\big(G(P, Q), f(Q)\big) = f(P) - \displaystyle\int_B f(q) \, \frac{\partial G(P, q)}{\partial n_q} \, dS_q,$

(b) $\quad E\big(N(P, Q), f(Q)\big) = f(P).$

Exercise 2. Define a Green's function for a non self-adjoint equation, and for the adjoint of this equation. How does the symmetry property (7.3.5) generalize to this case?

Exercise 3. If a Green's function for D exists, the solution of the Dirichlet problem for D is unique. *Hint:* Use (7.3.8).

Exercise 4. If the domain D increases, show that the Green's function $G(P, Q)$ also increases.

Exercise 5. If $q(P)$ increases, a solution $u(P)$ with $u(p) = f(p)$ decreases. *Hint:* An integral equation for $Lu(P) = \delta q(P)u(P)$.

Exercise 6. In Euclidean three-space, $\Delta u = k^2 u$ has separated solutions

$$r^{-\frac{1}{2}}Z_{s+\frac{1}{2}}(kr)P_s^m(\cos\theta)e^{\pm im\varphi},$$

which are single-valued if m and s are integers. Here $Z_{s+\frac{1}{2}}(x)$ is any solution of Bessel's equation of order $s+\frac{1}{2}$, while $P_s^m(\cos\theta)$ is the associated Legendre function.

7.4. The kernel function. We shall now introduce another domain functional which is related to the functions of Green and Neumann for a domain D, and yet has quite different properties. Let us consider the difference

$$K(P, Q) = N(P, Q) - G(P, Q) = n(P, Q) - g(P, Q), \qquad (7.4.1)$$

of these two fundamental solutions. This function $K(P, Q)$ is finite for $P = Q$, because the singularities of $N(P, Q)$ and $G(P, Q)$ cancel. Clearly $K(P, Q)$ is symmetric because both $N(P, Q)$ and $G(P, Q)$ are. Since the functions $n(P, Q)$ and $g(P, Q)$ are both solutions of the differential equation for P in D, and Q fixed, or Q in D, with P fixed, the same is true of $K(P, Q)$. Thus, for Q fixed in D, $K(P, Q)$ is a finite-valued solution function. On the boundary B, $N(p, Q)$ is non-negative while $G(p, Q)$ is zero; thus $K(p, Q)$ is non-negative on B. Therefore $K(P, Q)$ is non-negative in the interior domain D: $K(P, Q) \geqslant 0$.

This functional $K(P, Q)$ is known as the reproducing kernel of the domain D with respect to the differential equation $\Delta u = qu$, because of a property which we shall now establish. From (7.3.4) and (7.3.10) we see that if $u(P)$ is any solution function,

$$u(P) = \int_B \left(\frac{\partial N(P,q)}{\partial n} - \frac{\partial G(P,q)}{\partial n} \right) u(q)\, dS_q = \int_B \frac{\partial K(P,q)}{\partial n} u(q)\, dS_q.$$
(7.4.2)

Recalling that $K(P,Q)$, as function of Q, is a solution of the differential equation, we may apply Green's formula to (7.4.2) and so find

$$u(P) = E(K(P,Q), u(Q)),$$
(7.4.3)

where $E(K(P,Q), u(Q))$ is the Dirichlet integral whose argument functions are $K(P,Q)$ and $u(Q)$. Thus, the 'scalar product' of $u(Q)$ with the kernel $K(P,Q)$ in the Dirichlet integral reproduces just the solution function $u(P)$. This formula has some interesting consequences which we will shortly explore.

Quite apart from the property expressed by (7.4.3), we see that formula (7.4.2) provides another representation formula for the solution of the Dirichlet problem with boundary values $u(p) = f(p)$. Indeed, let $f(P)$ be any differentiable function with given boundary values, and let us form its Dirichlet integral with $K(P,Q)$. We see that

$$E(K(P,Q), f(Q))$$
$$= - \int_D f(Q)\{\Delta_Q K(P,Q) - q(Q)K(P,Q)\}\, dV_Q + \int_B f(q)\, \frac{\partial K(P,q)}{\partial n}\, dS_q,$$

and the volume integral on the right is zero because $K(P,Q)$ is a solution function. But by (7.4.2) we see that the surface integral, which, as a function of P, is a solution function, must have the boundary values $f(p)$. Thus the solution of the Dirichlet problem with data given by $f(p)$ may be written in either of the forms

$$u(P) = \int_B \frac{\partial K(P,q)}{\partial n} u(q)\, dS_q = E(K(P,Q), f(Q)).$$
(7.4.4)

Corresponding to (7.4.2), we have a representation formula for the Neumann problem. From (7.3.11), and the vanishing of $G(P,Q)$ for P on B, we see that for any solution function $u(P)$,

$$u(P) = \int_B \{N(P,q) - G(P,q)\} \frac{\partial u(q)}{\partial n}\, dS_q$$
$$= \int_B K(P,q) \frac{\partial u(q)}{\partial n}\, dS_q.$$
(7.4.5)

Thus a knowledge of $K(P, Q)$ enables us to write down a solution of the Neumann problem as well.

As an example of a kernel function, let us consider the equation $\Delta u = k^2 u$ ($k > 0$) in Euclidean three-space, and let us specify the region as the upper half-space $z > 0$. In § 7.2 we saw that this equation has the fundamental singularity $(4\pi r)^{-1}\exp[-kr]$, where $r = r(P, Q)$ is the

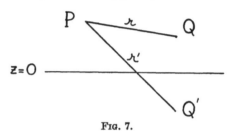

FIG. 7.

distance function. Let Q' be the reflection of any point Q in the plane $z = 0$, and let r' denote $r(P, Q') = r(P', Q)$ (Fig. 7). When Q lies on the boundary $z = 0$, Q' coincides with Q, r' coincides with r, and, as is easily verified,

$$\frac{\partial r'}{\partial n} = -\frac{\partial r}{\partial n},$$

when Q varies. From these observations it is easy to determine $G(P, Q)$ and $N(P, Q)$. Indeed, the Green's function, which is symmetric in P and Q and vanishes for Q on the boundary, is evidently given by

$$G(P, Q) = \frac{1}{4\pi}\left[\frac{e^{-kr}}{r} - \frac{e^{-kr'}}{r'}\right].$$

Similarly, we find that

$$N(P, Q) = \frac{1}{4\pi}\left[\frac{e^{-kr}}{r} + \frac{e^{-kr'}}{r'}\right]$$

has a vanishing normal derivative on the boundary. Subtracting, we conclude that

$$K(P, Q) = \frac{1}{2\pi}\frac{e^{-kr'}}{r'} \qquad (r' = r(P, Q')).$$

Note that $K(P, P)$ is finite unless P lies on the boundary. Since this kernel function is exponentially small when r (and therefore r') tends to infinity, integrals over an infinite domain as in (7.4.4) or (7.4.5) will converge.

The reproducing property (7.4.3) determines the kernel $K(P, Q)$ uniquely. For, if there were a different function $K_1(P, Q)$ with the same property, we could form the expression

$$E(K(P, Q), K_1(Q, R)).$$

Since $K(P, Q)$ reproduces, this Dirichlet integral is equal to $K_1(P, R)$, and since $K_1(P, Q)$ reproduces, it is equal to $K(P, R)$. Thus the two kernels are identical, and the uniqueness of $K(P, Q)$ is proved.

The Dirichlet integral is a positive definite bilinear functional, and so satisfies the Schwarz inequality (§ 4.3, Ex. 4):

$$|E(u, v)|^2 \leqslant E(u, u)E(v, v).$$

If we apply this inequality to the right-hand side of (7.4.3), we find

$$|u(P)|^2 = |E(K(P, Q), u(Q))|^2$$
$$\leqslant E(K(P, Q), K(P, Q))E(u(Q), u(Q))$$
$$= K(P, P)E(u, u).$$

Thus every solution satisfies an inequality of the form

$$|u(P)| \leqslant (K(P, P))^{\frac{1}{2}}(E(u, u))^{\frac{1}{2}}. \tag{7.4.6}$$

On the right are two factors, one depending only on the point P, the other on the solution $u(P)$ but not on P. The function $K(P, P)$ is finite in the interior of D but tends to infinity as P tends to a point of the boundary.

The Dirichlet integral $E(u, u)$ may be regarded as a measure of the total size or norm of the solution function $u(P)$ on the domain D. Let us consider the entire family F of solutions $u(P)$ on D whose Dirichlet integrals are less than or equal to a fixed constant:

$$E(u, u) \leqslant A^2. \tag{7.4.7}$$

We will show that the family F is 'compact', in this sense: that *any sequence* $\{u_n(P)\}$ $(n = 1, 2, ...)$ *of functions of F contains a subsequence which converges to a function again belonging to the family F.*

First, we see that if $u(P)$ is contained in F,

$$|u(P)| \leqslant (K(P, P))^{\frac{1}{2}}(E(u, u))^{\frac{1}{2}} \leqslant (K(P, P))^{\frac{1}{2}}A. \tag{7.4.8}$$

Thus functions of F are uniformly bounded in any fixed closed region D_1 interior to D, since $K(P, P)$ is bounded in any such region D_1. Next, let us differentiate (7.4.3) with respect to the coordinates of P, and then apply the Schwarz inequality. We find

$$\left|\frac{\partial u(P)}{\partial x^i}\right|^2 = \left|E\left(\frac{\partial K(P, Q)}{\partial x^i}, u(Q)\right)\right|^2$$
$$\leqslant E\left(\frac{\partial K(P, Q)}{\partial x^i}, \frac{\partial K(P, Q)}{\partial x^i}\right)E(u, u) \tag{7.4.9}$$
$$\leqslant E\left(\frac{\partial K(P, Q)}{\partial x^i}, \frac{\partial K(P, Q)}{\partial x^i}\right)A^2.$$

The Dirichlet integral on the right is again finite and bounded in the interior region D_1. It follows that the partial derivatives of $u(P)$ are uniformly bounded independently of $u(P)$. Thus the functions $u(P)$ of F are equicontinuous, uniformly in any closed region D_1 contained in the interior of D.

The first derivatives $\partial u/\partial x^i$ are also equicontinuous in D_1. This can be shown if we differentiate (7.4.3) twice with respect to the coordinates of P before applying the Schwarz inequality as in (7.4.9).

We shall make use of the following theorem of Ascoli: any infinite sequence of functions, uniformly bounded and equicontinuous on a bounded closed region, contains a subsequence which converges uniformly to a continuous limit. A proof may be found in (20, p. 265).

On the region D_1, the solution functions $u_n(P)$ of the given sequence are uniformly bounded and equicontinuous; we may apply the Ascoli theorem and select an infinite subsequence $\{u_n\}$ uniformly convergent in D. Disregarding from now on any other members of F, we select the derivatives $\partial u_n/\partial x^1$ which, by the above, are also uniformly bounded and equicontinuous in D. Hence, by Ascoli's theorem, there exists an infinite subsequence for which the $\partial u_n/\partial x^1$ are uniformly convergent in D. Continuing in this way we find that F contains an infinite sequence $\{u_n^1\}$ which, together with the sequences of first derivatives, is uniformly convergent in D_1.

Let the limit at P of the convergent sequence $\{u_n^1\}$ be denoted by $u^1(P)$. If S is a small sphere containing P, and $G(P, Q)$ the Green's function of S, we see that

$$u^1(P) = \lim_{m \to \infty} u_m^1(P) = -\lim_{m \to \infty} \int_S u_m^1(q) \frac{\partial G(P, q)}{\partial n_q} \, dS_q$$

$$= -\int_S [\lim_{m \to \infty} u_m^1(q)] \frac{\partial G(P, q)}{\partial n_q} \, dS_q,$$

since the convergence is uniform on S. Thus $u^1(P)$ must be a solution of the differential equation in the interior of D. A similar calculation shows that the derivatives $\partial u_n^1/\partial x^i$ tend to the limit $\partial u^1/\partial x^i$ as n tends to infinity.

If we denote by $E_1(u^1, u^1)$ the Dirichlet integral of u^1 over D_1, we see that

$$E_1(u^1, u^1) = \lim_{m \to \infty} E_1(u_n^1, u_n^1) \leqslant A^2$$

since the convergence of the u_n^1 and their first derivatives to u^1 and its first derivatives is uniform in D_1, and since the u_n^1 belong to F.

Now let D_k ($k = 1, 2, ...$) be a sequence of closed domains, each containing the preceding, which tend to D as k tends to infinity. Now the sequence u_n^1 is likewise uniformly bounded and equicontinuous in D_2. By the Ascoli theorem, $\{u_n^1\}$ contains a subsequence $\{u_n^2\}$ which, together with the derivative subsequence, converges uniformly in D_2 to a limit $u^2(P)$. Clearly $u^2(P)$ is a solution function; and $u^2(P) = u^1(P)$ in D_1. Also

$$E_2(u^2, u^2) = \lim_{n \to \infty} E_2(u_n^2, u_n^2) \leqslant A^2,$$

and since D_1 is contained in D_2 we have

$$E_1(u^1, u^1) \leqslant E_2(u^2, u^2).$$

Repeating this process for each domain D_k, we construct a sequence of solution functions $\{u_n^k\}$, having as limit a solution u^k defined in D_k. Also, $u^k = u^{k-1}$ in D_{k-1}, so that the definition

$$u(P) = u^k(P)$$

for P in D_k, is consistent, and defines a single solution function in the whole interior of D.

The rest of the convergence proof now follows quickly. As definitive subsequence of the original we pick the diagonal sequence u_k^k. This infinite subsequence converges to $u(P)$ uniformly in each region D_k. Finally, $u(P)$ belongs to F, because

$$E(u, u) = \lim_{k \to \infty} E_k(u^k, u^k) \leqslant A^2.$$

Here the limit exists since the $E_k(u^k, u^k)$ form an increasing sequence bounded above by A^2.

In this proof we have not assumed that the solutions are finite on the boundary of D. We also see that the convergence to the limit is uniform in any closed interior region, since any such region lies inside the D_k for k sufficiently large. The solutions of an elliptic differential equation form a special class having convergence properties which wider classes of real-valued functions fail to possess.

Exercise 1. Show that $K(P, Q)^2 \leqslant K(P, P)K(Q, Q)$.

Exercise 2. Verify the following relations connecting the functions $K(P, Q)$, $G(P, Q)$, $N(P, Q)$, and any fundamental solution $\gamma(P, Q)$:

$$G(P, Q) = \gamma(P, Q) - \int_B \frac{\partial K(P, r)}{\partial n_r} \gamma(r, Q) \, dS_r,$$

$$N(P, Q) = \gamma(P, Q) - \int_B K(P, r) \frac{\partial \gamma(r, Q)}{\partial n_r} \, dS_r,$$

$$K(P, Q) = - \int_B N(P, r) \frac{\partial G(r, Q)}{\partial n_r} \, dS_r.$$

Exercise 3. If $R(P, Q)$ is a Robin function for D, show that the kernel $K_h(P, Q) = R(P, Q) - G(P, Q)$ has the reproducing property with respect to the modified Dirichlet integral

$$E(u, v) + \int_B huv \, dS.$$

Exercise 4. Show that the family of solutions $u(P)$ which are uniformly bounded in D: $|u(P)| \leqslant A$ are also (a) equicontinuous in D; (b) compact. *Hint*: Use (7.3.4).

Exercise 5. If a sequence of solution functions $u_n(P)$ converges uniformly on the boundary B, then it converges uniformly in the interior, together with the sequences of derivatives of all orders; and the limit is again a solution function.

VIII

EIGENFUNCTIONS

As yet, we have studied boundary value problems for elliptic equations only under a condition on the coefficient $q(P)$ which ensures that the solutions are unique. In particular, the Laplace equation $\Delta u = 0$ has been excluded by this restriction. We shall now relax this condition, and study the self-adjoint equation

$$\Delta u(P) + c(P)u(P) = 0,$$

without restricting the sign of the coefficient $c(P)$.

Non-uniqueness of a non-homogeneous problem implies the existence of a solution of the corresponding homogeneous problem, that is, of an eigenfunction. These special solutions are important quite apart from their connexion with boundary value problems.

First we consider Laplace's equation and modify the work of Chapter VII in order to prove the standard boundary value theorems for harmonic functions, and to construct the functions of Green and Neumann. Turning to the general self-adjoint equation, we study the Poisson equation in a closed space, and on a domain with boundary. In the last two sections we consider eigenvalues and eigenfunctions in their own right and show how they may be applied to initial value problems for parabolic and hyperbolic equations.

8.1. Harmonic functions. The great historical and physical importance of Laplace's equation $\Delta u = 0$ has resulted in a detailed study of harmonic functions, especially in Euclidean three-space. We shall not here attempt to discuss special types of solutions such as spherical harmonics; these have been treated in another volume of this series (29). Rather, we shall adapt the methods of the preceding chapter to the Laplace equation, and will establish boundary value theorems for harmonic functions.

As in § 7.1, let D be a domain with boundary B, and let F be the double, a closed Riemannian space. The equation

$$Lu(P) = \Delta u(P) - q(P)u(P) = 0 \qquad (8.1.1)$$

has a Green's function $g(P, Q)$ in F provided that $q(P) \geqslant 0$, $q(P) \not\equiv 0$.

Let us now suppose that

$$q(P) = 0 \quad (\text{in } D)$$

$$q(P) > 0 \quad (\text{in } F-D). \tag{8.1.2}$$

Then all solutions of (8.1.1) are harmonic functions in D, and there exists a Green's form $g(P, Q)$ in F for (8.1.1). Thus

$$\Delta_P g(P, Q) = 0, \tag{8.1.3}$$

for P in D, $P \neq Q$; and $g(P, Q)$ is symmetric in P and Q.

Let us now examine the boundary value theorems of § 7.2, taking first the Dirichlet problem for D. To establish the result for harmonic functions, we may again write the integral equation (7.2.3) with $\lambda = +1$. There exists a solution if and only if the datum function $f(p)$ is orthogonal to every solution of the homogeneous transposed equation (7.2.5). As in § 7.2, if (7.2.5) has a non-trivial eigenfunction, there exists a single layer potential (7.2.6) satisfying (7.2.7). But, according to (8.1.2), the Neumann problem for (8.1.1) in $\tilde{D} = F-D$ has at most one solution. Thus $V(P)$ must be zero in $\tilde{D}$, and, being continuous across B, must have vanishing boundary values in D. We recall from § 5.1 that the Dirichlet problem for harmonic functions has at most one solution; hence $V(P)$ is identically zero in D. It then follows from the discontinuity conditions that the eigenfunction of (7.2.5) is identically zero. Hence (7.2.3) has always a solution in this case. *The Dirichlet problem for harmonic functions on D has a unique solution.*

For the Neumann problem on D, we must satisfy the necessary condition (5.1.9), which states that the average assigned value of the normal derivative shall be zero. The integral equation for the interior Neumann problem is, in the notation of § 7.1,

$$g(p) = \psi(p) + \int_B k(q, p)\psi(q) \, dS_q, \tag{8.1.4}$$

where

$$k(q, p) = 2\frac{\partial\gamma(q, p)}{\partial n_p}.$$

If the homogeneous transposed equation

$$0 = \chi(p) + \int_B k(q, p)\chi(q) \, dS_q \tag{8.1.5}$$

has a non-zero solution $\chi(p)$, then the double layer potential

$$W(P) = \int_B \chi(q) \frac{\partial\gamma(P, q)}{\partial n_q} \, dS_q$$

satisfies the boundary condition

$$W_+(p) = W(p) + \tfrac{1}{2}\chi(p) = 0.$$

Since $W(P)$ satisfies $LW(P) = 0$ in $\check{D}$, and vanishes on the boundary, $W(P) \equiv 0$ in $\check{D}$. The normal derivative being continuous across B, we have

$$\frac{\partial W_-(p)}{\partial n_p} = \frac{\partial W_+(p)}{\partial n_p} = 0.$$

That is, in D, $W(P)$ is a harmonic function whose normal derivative on B is zero. Thus

$$W(P) = \text{constant} \quad (P \text{ in } D).$$

Finally, by the discontinuity conditions (5.4.14),

$$\chi(p) = W_+(p) - W_-(p) = \text{constant.}$$

Thus $\chi(p) = \text{constant}$ is the only non-zero solution of (8.1.5). Hence (8.1.4) has a solution if $g(p)$ is orthogonal to 1 on B. It follows that *there exists a solution of the Neumann problem for $\Delta u = 0$ in D:*

$$\frac{\partial u}{\partial n} = g(p)$$

if and only if
$$\int_B g(p)\, dS_p = 0. \qquad (8.1.6)$$

Thus the necessary condition is also sufficient. To any solution of the Neumann problem we may add an arbitrary constant which contributes nothing to the normal derivative. Indeed, the additive constant in the solution $u(P)$ may be chosen so that the average value of $u(P)$ itself on B is zero:

$$\int_B u(p)\, dS_p = 0. \qquad (8.1.7)$$

When this condition is applied, $u(P)$ is uniquely determined.

The Robin problem for harmonic functions on D can be discussed exactly as in § 7.2, and the existence proof given there is valid in this case, as the reader may verify. Thus there exists a unique harmonic function satisfying a given Robin boundary condition. For three-dimensional Euclidean space the results of this section are the boundary value theorems of Newtonian potential theory.

In the Euclidean plane, where the study of harmonic functions is related to complex function theory, there is possible a somewhat different formulation of the Dirichlet and Neumann boundary value problems, in which the nature of the condition (8.1.6) is further revealed. If

$$f(z) = f(x+iy) = u(x,y) + iv(x,y),$$

then the conjugate harmonic functions u, v satisfy the Cauchy–Riemann equations

$$\frac{\partial u}{\partial x} = \frac{\partial v}{\partial y}, \qquad \frac{\partial u}{\partial y} = -\frac{\partial v}{\partial x}.$$

In symbolic notation we see that $(\nabla u)^2 = (\nabla v)^2$, $\nabla u . \nabla v = 0$ (Fig. 8.) Let C be a closed analytic curve bounding a domain D, and let s, n measure

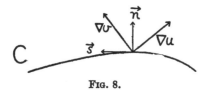

FIG. 8.

arc length counterclockwise along C, and along the outward normal to C, respectively. Thus

$$\frac{\partial u}{\partial s} = \frac{\partial u}{\partial x}\frac{\partial x}{\partial s} + \frac{\partial u}{\partial y}\frac{\partial y}{\partial s} = \frac{\partial v}{\partial y}\left(-\frac{\partial y}{\partial n}\right) - \frac{\partial v}{\partial x}\frac{\partial x}{\partial n} = -\frac{\partial v}{\partial n},$$

and similarly,
$$\frac{\partial u}{\partial n} = \frac{\partial v}{\partial s}.$$

We now remark that the Dirichlet boundary condition may be expressed as follows: if $f(s)$ is a given function on C, let $\partial u/\partial s = f(s)$ on C, and let the value of u at a single point $s = 0$ on C be given. Thus

$$u(s) = u(0) + \int_0^s f(s)\, ds$$

is determined at every point of the boundary. But on integrating around the entire curve C, we must regain for u the value $u(0)$. Thus we must postulate

$$\oint f(s)\, ds = 0.$$

This condition on the data of the Dirichlet problem, stated in this form, is necessary to ensure that a single-valued solution shall result. When the condition holds, the Dirichlet problem has a well-determined solution $u(P) = u(x, y)$. Let $v(x, y)$ be the harmonic conjugate of u, determined up to an additive constant by the path-independent line integral

$$v(x, y) = \int \left[\frac{\partial u}{\partial x}\, dy - \frac{\partial u}{\partial y}\, dx\right].$$

Then we see that $\partial v/\partial n = f(s)$. *The Dirichlet problem for $u(x, y)$ is equivalent to the Neumann problem for the harmonic conjugate function $v(x, y)$.* Stated in this way, the solution of either problem is undetermined by an additive constant, which may be specified by giving the function at a point of C. The necessary condition that the datum function $f(s)$ should have a zero average value on S appears in Dirichlet's problem as a condition of single-valuedness, and in Neumann's problem as a consequence of Green's formula.

Exercise 1. Show that the conductor potential

$$W(P) = \int_B \frac{\partial \gamma(P,q)}{\partial n_q}\, dS_q$$

is equal to zero in $F-D$ and unity in D.

8.2. Harmonic domain functionals. The Green's function $g(P,Q)$ of the preceding section satisfies the Laplace equation in the domain D and is in consequence a fundamental solution in the large for Laplace's equation in D. With the help of this fundamental singularity we can construct the Green's function of the domain D itself. Since the Dirichlet problem on D has always a solution, there exists for each P a harmonic function of Q in D with boundary values $-g(P,q)$. Adding this function to $g(P,Q)$ we obtain the Green's function $G(P,Q)$ for Laplace's equation in D. The properties of $G(P,Q)$ can be worked out exactly as in § 7.3, except for the proof that $G(P,Q)$ is non-negative in D, and this can be shown by the method used for the Neumann function in that section. The verification will be left as an exercise.

The harmonic Green's function for the interior of a sphere in Euclidean space may be found explicitly by the use of points which are inverse with respect to the sphere. This method appears to succeed only for harmonic functions in Euclidean space, but is applicable in any number N of dimen-

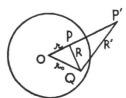

FIG. 9. Inverse points.

sions. Let the origin O be the centre, a the radius of the sphere S (Fig. 9). If P is any point within S, its inverse point P' lies on the line OP, and at a distance OP' such that $OP'.OP = a^2$. Let OP have length r, then OP' has length a^2/r. If now Q is any argument point within S, and $PQ = R$, $P'Q = R'$, we know that, as functions of Q, the quantities R^{-N+2}, R'^{-N+2}

are harmonic, the former only having a singularity within S. Also, when Q lies on the surface of the sphere, we have

$$\frac{R'}{R} = \frac{a}{r},$$ (8.2.1)

since the triangles OPQ, OQP' are then similar. That is, the harmonic function

$$\frac{1}{R^{N-2}} - \left(\frac{a}{r}\right)^{N-2} \frac{1}{R'^{N-2}}$$

vanishes for Q on the surface of the sphere, and is infinite like R^{-N+2} when Q tends to P. Thus the Green's function is

$$G(P,Q) = \frac{1}{\omega_N(N-2)}\left[\frac{1}{R^{N-2}} - \left(\frac{a}{r}\right)^{N-2} \frac{1}{R'^{N-2}}\right].$$ (8.2.2)

If P and Q subtend at the origin an angle θ, and $OP = r$, $OQ = r_0$, it is easily found that

$$R^2 = r_0^2 + r^2 - 2rr_0\cos\theta,$$

$$R'^2 = r_0^2 + a^4/r^2 - 2a^2(r_0/r)\cos\theta.$$ (8.2.3)

Then, since the outward normal to the sphere is in the direction of increasing r_0, we find, making use of the fact that $rR' = aR$ on S,

$$\frac{\partial G(P,q)}{\partial n_q} = \frac{\partial G(P,q)}{\partial r_0}\bigg]_{r_0=a} = -\frac{a^2-r^2}{a\omega_N R^N}.$$ (8.2.4)

Therefore the representation formula for a harmonic function which takes given values $f(p)$ on the sphere $r = a$ is

$$u(P) = \frac{1}{\omega_N}\int_S f(q)\frac{a^2-r^2}{[a^2+r^2-2ar\cos\theta]^{\frac{1}{2}N}}\,dS.$$ (8.2.5)

Here θ is the angle subtended at the origin by the point P and the variable point q. This is the Poisson integral formula.

In this example we must suppose $N > 2$; as an exercise for the reader we suggest the similar construction in two dimensions when the final form of the Poisson formula is given in (1.2.6).

In the general case, the definition of the Neumann function needs explicit modification, compared with the definition given in § 7.3, since we can no longer require that the normal derivative of $N(P,Q)$ should vanish on B. For, if $\gamma(P,Q)$ is any fundamental singularity, we have from Green's formula (where Q is the variable integrated)

$$E_{D-S_\epsilon(P)}(\gamma(P,Q),1) = \int_{B-bS_\epsilon(P)} \frac{\partial\gamma(P,q)}{\partial n}\,dS_q,$$ (8.2.6)

and since the limit as $\epsilon \to 0$ of the surface integral over the small sphere about P is -1, we must have

$$\int_B \frac{\partial \gamma(P,q)}{\partial n} \, dS_q = -1. \tag{8.2.7}$$

In view of (8.2.7) the simplest condition which can be made on the normal derivative of the Neumann function is

$$\frac{\partial N(P,q)}{\partial n_q} = -c, \tag{8.2.8}$$

where c is the reciprocal of the surface area $\int_B dS$. To satisfy (8.2.8), let

$$N(P,Q) = \gamma(P,Q) + n(P,Q), \tag{8.2.9}$$

where $n(P,Q)$ is that harmonic function of Q, with

$$\frac{\partial n(P,q)}{\partial n} = -\frac{\partial \gamma(P,q)}{\partial n} - c. \tag{8.2.10}$$

In view of (8.2.7) the condition (8.1.6) is satisfied, so that $n(P,Q)$ exists, and is determined up to an additive constant. Thus in (8.2.9), $N(P,Q)$ is determined up to an additive constant. Let us choose this constant so that

$$\int_B N(P,q) \, dS_q = 0, \tag{8.2.11}$$

thus determining $N(P,Q)$ uniquely. The symmetry property

$$N(P,Q) = N(Q,P) \tag{8.2.12}$$

may now be proved by considering the integral analogous to $I(\epsilon)$ in § 7.3, and using (8.2.8) and (8.2.11). This proof we shall again leave as an exercise. Since $N(P,Q)$ is symmetric, and of the form (8.2.9), it is a fundamental singularity. The representation formula for the Neumann problem may be found by inserting $N(P,Q)$ in (5.4.6), which yields

$$u(P) = \int_B \left(N(P,q) \frac{\partial u(q)}{\partial n} - u(q) \frac{\partial N(P,q)}{\partial n} \right) dS_q$$

$$= \int_B N(P,q) \frac{\partial u(q)}{\partial n} \, dS_q + c \int_B u(q) \, dS_q. \tag{8.2.13}$$

Thus, if we impose on $u(P)$ the condition (8.1.7), the second term vanishes, and the representation formula is formally the same as in § 7.3.

Again, the difference

$$K(P,Q) = N(P,Q) - G(P,Q) \tag{8.2.14}$$

is a reproducing kernel for those harmonic functions which satisfy (8.1.7). Indeed, since $K(P, Q)$ is in each variable a regular harmonic function in D we have from (8.2.13) for any harmonic function $u(P)$:

$$u(P) = \int_B K(P, q) \frac{\partial u(q)}{\partial n} \, dS_q + c \int_B u(q) \, dS_q$$

$$= E(K(P, Q), u(Q)) + c \int_B u(q) \, dS_q. \qquad (8.2.15)$$

If (8.1.7) holds, the last term is zero, and the reproducing property of $K(P, Q)$ is established. Thus the formal properties of § 7.3 extend to the harmonic reproducing kernel.

The convergence theorem of § 7.4 may also be applied to harmonic functions which satisfy (8.1.7); in fact the proofs are valid without alteration, once the reproducing property is proved. However the condition (8.1.7), or one equivalent to it, is necessary for this purpose because, for example, the sequence $u_n(P) = n$ is a sequence of harmonic functions with zero Dirichlet integrals which does not converge at any point.

Exercise 1. In the Euclidean plane, let D be a simply-connected domain, and Q a point within D. Let $\zeta = f(z)$ map D, in the z-plane, into the unit circle in the ζ-plane, with Q being mapped upon the origin $\zeta = 0$. Show that the Green's function $G(P, Q)$ of D for $u_{xx} + u_{yy} = 0$ is

$$G(P, Q) = -(1/2\pi) \log |f(x+iy)|.$$

Exercise 2. For the circular domain $r \leqslant a$ in the Euclidean plane, and with the notation of inverse points as above, show that

$$G(P, Q) = \frac{1}{2\pi} \log \frac{r' r_Q}{ra}, \qquad N(P, Q) = \frac{1}{2\pi} \log \frac{a^3}{rr' r_Q},$$

and calculate the kernel function $K(P, Q)$.

Exercise 3. If $G(P, Q)$ is the harmonic Green's function for an N-dimensional domain D, show that $\partial G/\partial n_q \leqslant 0$ on the boundary B. Hence show that a harmonic function non-negative on B is non-negative in D. Also show that a harmonic function has no proper maximum or minimum in the interior of its region of definition.

Exercise 4. Express in terms of the harmonic Green's function a solution of $\Delta u(P) = \rho(P)$ in D which takes boundary values $f(p)$ on B. *Hint*: See (7.3.6). Show that a solution of the problem exists.

8.3. The Poisson equation in a closed space. The proof of § 7.1 for the existence of a fundamental singularity in the large in a closed space was valid for the self-adjoint equation

$$Lu(P) = \Delta u(P) - q(P)u(P) = 0 \qquad (8.3.1)$$

provided that the coefficient $q(P)$ satisfied

$$q(P) \geqslant 0, \qquad q(P) \not\equiv 0. \qquad (8.3.2)$$

These conditions ensured that no non-trivial eigenfunctions of (8.3.1) exist in the closed space. We shall now relax this limitation and study (8.3.1) without restriction on the sign of the coefficient $q(P)$. It would be possible to apply the method of § 7.1 with some modifications. However, we shall follow a different approach.

Let us write the Poisson equation in the form

$$\Delta u(P) + c(P)u(P) = \rho(P), \qquad (8.3.3)$$

and let $q(P)$ be any smooth bounded function satisfying (8.3.2). If the differential operator L is defined as in (8.3.1), we may write (8.3.3) in the form

$$Lu(P) = q_1(P)u(P) + \rho(P), \qquad (8.3.4)$$

where

$$q_1(P) = -c(P) - q(P). \qquad (8.3.5)$$

Let us now regard the right-hand side of (8.3.5) as a volume density, and let us write down the solution of that equation using the Green's function $\gamma(P, Q)$ of § 7.2, pertaining to the operator L. Thus we find

$$u(P) = - \int_F \gamma(P, Q)[q_1(Q)u(Q) + \rho(Q)] \, dV_Q. \qquad (8.3.6)$$

Let us define

$$v(P) = - \int_F \gamma(P, Q)\rho(Q) \, dV_Q; \qquad (8.3.7)$$

this function is the unique solution in F of the equation

$$Lv(P) = \rho(P). \qquad (8.3.8)$$

Now (8.3.6) may be written as a Fredholm equation for the unknown function $u(P)$:

$$u(P) + \int_F \gamma(P, Q)q_1(Q)u(Q) \, dV_Q = v(P). \qquad (8.3.9)$$

The domain of integration is the whole space F. Conversely we see that any twice-differentiable solution of this integral equation provides a solution of the given differential equation. Indeed, if we apply the operator L to (8.3.9) and take account of (8.3.8), we find (8.3.4) which is just equivalent to (8.3.3). Thus the integral equation is equivalent to the differential equation.

Now (8.3.9) has as kernel the function $\gamma(P, Q)q_1(Q)$ which is regular and bounded except when $P \to Q$ when its singularity is of order s^{-N+2}. The first iterated kernel must then, according to § 6.4, have a singularity of order s^{-N+4}; and clearly the iterated kernels of sufficiently high order are continuous. We may therefore apply the Fredholm orthogonality theorem to (8.3.9).

The homogeneous transposed equation is

$$w(P) + q_1(P) \int_F \gamma(P, Q)w(Q)\, dV_Q = 0, \qquad (8.3.10)$$

since the factor $q_1(P)$ can be taken outside the integral sign, and since $\gamma(P, Q)$ is symmetric. There are at most a finite number of linearly independent solutions of (8.3.10), the number of these being moreover equal to the number of linearly independent solutions of the untransposed homogeneous equation

$$z(P) + \int_F \gamma(P, Q)q_1(Q)z(Q)\, dV_Q = 0. \qquad (8.3.11)$$

If we multiply (8.3.11) by $q_1(P)$, we find just the statement that $q_1(P)z(P)$ is a solution of (8.3.10). It follows that any solution $w_\mu(P)$ of (8.3.10) can be expressed in the form

$$w_\mu(P) = q_1(P)z_\mu(P),$$

where $z_\mu(P)$ is a solution of (8.3.11). We also note that the $z_\mu(P)$, being solutions of (8.3.11), satisfy the differential equation

$$Lz_\mu - q_1 z_\mu = \Delta z_\mu + cz_\mu = 0$$

on F; that is, they are just the eigenfunctions of the differential equation on F.

The Fredholm orthogonality condition for (8.3.9) may now be expressed as the vanishing of the integral

$$\int_F v(P)w_\mu(P)\, dV_P = \int_F v(P)q_1(P)z_\mu(P)\, dV_P = \int_F v(P)Lz_\mu(P)\, dV_P,$$

since $z_\mu(P)$ satisfies $Lz_\mu = q_1 z_\mu$. Applying Green's formula to this last integral we find for it the expression

$$\int_F z_\mu(P)Lv(P)\, dV_P,$$

and, in view of (8.3.8), the orthogonality condition finally becomes

$$\int_F z_\mu(P)\rho(P)\, dV_P = 0. \qquad (8.3.12)$$

When (8.3.12) holds, the integral equation has a solution of the form

$$u(P) = v(P) + \int_F R(P, Q)v(Q) \, dV_Q,$$

where $R(P, Q)$ is the appropriate resolvent kernel. To show that $u(P)$ satisfies the differential equation it is enough to show that $\Delta u(P)$ exists. By (8.3.8) it is clear that $\Delta v(P)$ exists. Since $\gamma(P, Q)$ is twice differentiable for $P \neq Q$, the same holds for the resolvent kernel. Under these circumstances a calculation of the type leading to (5.3.11) shows that the Laplacian operator can be applied to the above integral. Thus $u(P)$ possesses the requisite properties of regularity.

There exists a solution of $\Delta u + cu = \rho(P)$ *on* F *if and only if* $\rho(P)$ *is orthogonal on* F *to every solution* $z_\mu(P)$ *of*

$$\Delta z + cz = 0.$$

We observe that the most general solution of the given equation is

$$u(P) = u_0(P) + \sum_\mu c_\mu z_\mu(P),$$

where $u_0(P)$ is any particular solution. Since the $z_\mu(P)$ are linearly independent the constants c_μ may be chosen so that

$$\int_F u(P)z_\mu(P) \, dV_P = 0. \qquad (8.3.13)$$

If this orthogonality condition is included in the formulation of the problem, the solution is uniquely determined.

If the homogeneous equation $\Delta u + cu = 0$ has only the zero solution, that is, if there are no eigenfunctions, then the necessary condition is automatically satisfied and a solution of $\Delta u + cu = \rho$ always exists. Corresponding to each eigenfunction $z_\mu(P)$ there is a certain necessary condition (8.3.12), and also an arbitrary constant c_μ in the most general solution. The analogy with the theory of linear algebraic equations, and with the integral equation theory itself, is evident. We also see that the result is quite independent of the auxiliary operator L which figured in the proof.

Let the eigenfunctions $z_\mu(P)$ of $\Delta u + cu = 0$ in F be orthogonalized and normalized so that

$$\int z_\mu(P)z_\nu(P) \, dV_P = \delta_{\mu\nu}. \qquad (8.3.14)$$

Then, for an arbitrary function $f(P)$ on F, define the Fourier coefficients of $f(P)$ with respect to $z_\mu(P)$ as

$$f_\mu = \int_F z_\mu(P)f(P) \, dV_P. \qquad (8.3.15)$$

Then the right-hand side of

$$\Delta u(P)+c(P)u(P) = \rho(P)- \sum_{\mu} \rho_{\mu} z_{\mu}(P) \qquad (8.3.16)$$

is orthogonal to all eigenfunctions $z_{\mu}(P)$, by construction. Thus a solution of (8.3.16) exists, for arbitrary $\rho(P)$. Further, there exists a unique solution $u(P)$ which is itself orthogonal to the $z_{\mu}(P)$ as in (8.3.13).

This function $u(P)$ appears as a solution of the integral equation (8.3.9) in which $v(P)$ is expressed as an integral or volume potential of $\rho(P)$ over the domain F. Hence $u(P)$ itself can be expressed as an integral over F of the values of $f(P)$, which we shall write in the form

$$u(P) = - \int_{F} g(P, Q)\rho(Q)\, dV_{Q}, \qquad (8.3.17)$$

thus defining the Green's kernel $g(P, Q)$. A detailed examination of the way in which (8.3.17) was derived would show that $g(P, Q)$ is regular except when $P \to Q$ when it has a singularity asymptotic to $\gamma(P, Q)$.

Yet if one or more eigenfunctions exist, $g(P, Q)$ is not a fundamental singularity; since, if it were, (8.3.3) would be solvable for all $\rho(P)$ which is not the case. However, $g(P, Q)$ satisfies a related differential equation which may be found by substituting (8.3.17) into the equation (8.3.16). We find that this equation becomes

$$\Delta u+cu = -\{\Delta+c\} \int_{F} g(P, Q)\rho(Q)\, dV_{Q}$$

$$= \rho(P)- \int_{F} \{\Delta_{P}+c\}g(P, Q)\rho(Q)\, dV_{Q}$$

$$= \rho(P)- \sum_{\mu} \rho_{\mu} z_{\mu}(P).$$

The term $\rho(P)$ in the second line arises as in (5.3.11) from the singularity of $g(P, Q)$ as $Q \to P$. Cancelling the terms $\rho(P)$ and using (8.3.15), we have

$$\int_{F} [\{\Delta_{P}+c(P)\}g(P, Q)- \sum_{\mu} z_{\mu}(P)z_{\mu}(Q)]\rho(Q)\, dV_{Q} = 0, \qquad (8.3.18)$$

and this holds for arbitrary (suitably differentiable) functions $\rho(Q)$. In particular, let

$$\rho(Q) = s^{N}(P, Q)[\{\Delta_{P}+c(P)\}g(P, Q)- \sum_{\mu} z_{\mu}(P)z_{\mu}(Q)],$$

the factor s^{N} being inserted to ensure convergence of the integral. We have then a statement of the form

$$\int_{F} s^{N}(P, Q)k^{2}(Q)\, dV_{Q} = 0,$$

and so conclude that

$$k(Q) \equiv \{\Delta_P + c(P)\}g(P, Q) - \sum_\mu z_\mu(P)z_\mu(Q) = 0, \qquad Q \neq P.$$

Thus $g(P, Q)$ satisfies the equation

$$\{\Delta_P + c(P)\}g(P, Q) = \sum_\mu z_\mu(P)z_\mu(Q), \qquad P \neq Q. \tag{8.3.19}$$

The non-homogeneous term on the right is just the reproducing kernel in the L^2 metric for the eigenfunctions of $\Delta u + cu = 0$ on F. Indeed, denoting this kernel by $k(P, Q)$, and any eigenfunction by $z(P)$, we have

$$z(P) = \sum z_\mu z_\mu(P),$$

and
$$\int_F z(Q)k(P, Q)\, dV_Q = \int_F z(Q) \sum_\mu z_\mu(P)z_\mu(Q)\, dV_Q$$

$$= \sum_\mu z_\mu z_\mu(P) = z(P).$$

Since (8.3.17) is orthogonal to the $z_\mu(P)$, we have

$$0 = \int_F z_\mu(P)u(P)\, dV_P = -\int_F z_\mu(P)\, dV_P \int_F g(P, Q)\rho(Q)\, dV_Q$$

$$= -\int_F \rho(Q)\, dV_Q \int_F z_\mu(P)g(P, Q)\, dV_P,$$

and this holds for arbitrary $\rho(P)$. Hence

$$\int z_\mu(P)g(P, Q)\, dV_P = 0. \tag{8.3.20}$$

The symmetry property $g(P, Q) = g(Q, P)$ may now be established as in § 7.1.

In the special case of Laplace's equation $(c(P) \equiv 0)$, the kernel $k(P, Q)$ has an especially simple form, since $z(P) = $ constant is the only eigenfunction. If V is the volume $\int_F dV$ of F, the normalized eigenfunction is $V^{-\frac{1}{2}}$. Thus the harmonic Green's function of F satisfies

$$\Delta g(P, Q) = \frac{1}{V}.$$

An example of a harmonic Green's function of this kind is that of the unit sphere $r = 1$ in Euclidean three-space. In spherical polars we have, on the surface,
$$ds^2 = d\theta^2 + \sin^2\theta\, d\varphi^2. \tag{8.3.21}$$

Thus we consider the equation

$$\Delta u = \frac{1}{\sin\theta} \frac{\partial}{\partial\theta}\left(\sin\theta\, \frac{\partial u}{\partial\theta}\right) + \frac{1}{\sin^2\theta} \frac{\partial^2 u}{\partial\varphi^2} = \frac{1}{4\pi}.$$

A solution having a logarithmic singularity at the pole $\theta = 0$ is $-(2\pi)^{-1}\log(2\sin\tfrac{1}{2}\theta)$. If the pole is taken at an arbitrary point Q, then θ is the angle of arc from P to Q and so is equal to the geodesic distance $s(P,Q)$. The Green's function is then

$$g(P,Q) = -\frac{1}{2\pi}\log(2\sin\tfrac{1}{2}s). \tag{8.3.22}$$

Exercise 1. If $w_\mu(P)$ is a given solution of (8.3.10), find explicitly a function $z_\mu(P)$ with $q_1(P)z_\mu(P) = w_\mu(P)$, and show that this $z_\mu(P)$ also satisfies (8.3.11).

Exercise 2. Show that the c_μ may be chosen so that (8.3.13) holds, by solving a certain set of linear algebraic equations whose determinant is not zero.

Exercise 3. If $g(P,Q)$ is the harmonic Green's function show that $g(P,Q)-g(P,R)-g(Q,R)$ is a harmonic fundamental singularity in any region not containing the fixed point R.

Exercise 4. Show that $-(2\pi)^{-1}\log(\tan\tfrac{1}{2}s)$ is harmonic on the two-dimensional unit sphere, and discuss its singularities.

Exercise 5. In a closed N-space of constant positive curvature R, show that a harmonic Green's function is

$$g(P,Q) = -\frac{1}{\omega_{N+1}}\int_{b_N}^{s/R}\operatorname{cosec}^{N-1}t\,dt\int_\pi^t\sin^{N-1}v\,dv,$$

where $s = s(P,Q)$ and $0 < b_N < \pi$.

8.4. Dirichlet's problem for the Poisson equation. In the preceding section we saw that the solution of the Poisson equation

$$\Delta u(P)+c(P)u(P) = \rho(P) \tag{8.4.1}$$

in a closed space F, could be found by solving an integral equation with domain F, the kernel of which was related to the Green's function for a certain auxiliary operator. Let us apply this method to the same equation (8.4.1) on a domain D with boundary B, and let us attempt to find a solution function $u(P)$ taking assigned boundary values $f(p)$ on B.

For this problem we may take as our auxiliary operator the Laplacian Δ itself. From § 8.2 we recall that a harmonic Green's function $G(P,Q)$ for D exists, and, as in § 7.3, we observe that the volume potential

$$-\int_D G(P,Q)\rho(Q)\,dV_Q$$

is the solution of $\Delta u(P) = \rho(P)$ which vanishes on the boundary. Thus, if we denote by $v_0(P)$ the harmonic function with boundary values $v_0(p) = f(p)$, then the solution of $\Delta u(P) = \rho(P)$ with these same boundary values is evidently

$$u(P) = v_0(P) - \int_D G(P, Q)\rho(Q)\, dV_Q. \tag{8.4.2}$$

Let us now rewrite (8.1.1) in the form

$$\Delta u(P) = \rho(P) - c(P)u(P), \tag{8.4.3}$$

and regard the right-hand side as a volume density. Our problem is to find the solution of (8.4.3) with boundary values $f(p)$. But, by (8.4.2), this solution must satisfy

$$u(P) = -\int_D G(P, Q)[\rho(Q) - c(Q)u(Q)]\, dV_Q + v_0(P). \tag{8.4.4}$$

This is again an integral equation over the domain D. Transposing to the left all terms containing the unknown function $u(P)$, we find

$$u(P) - \int_D G(P, Q)c(Q)u(Q)\, dV_Q = v(P), \tag{8.4.5}$$

where the new function

$$v(P) = v_0(P) - \int_D G(P, Q)\rho(Q)\, dV_Q \tag{8.4.6}$$

comprises the non-homogeneous terms in (8.4.4). According to (8.4.2), we see that $v(P)$ satisfies the two conditions

$$\Delta v(P) = \rho(P), \qquad v(p) = v_0(p) = f(p), \tag{8.4.7}$$

and it is clear that $v(P)$ is uniquely determined by these conditions.

Now any twice-differentiable solution of (8.4.5) is a solution of the differential equation (8.4.1), as we see by operating on (8.4.5) with the Laplacian, and noting the first of (8.4.7). Likewise, any solution of (8.4.5) takes on the boundary values of $v(P)$ which are equal to the required boundary values $f(p)$. Thus the problem now is to find a twice-differentiable solution of (8.4.5).

The kernel of this integral equation is $c(Q)G(P, Q)$, where $c(Q)$ is assumed to be bounded on D. Since the Green's function vanishes on the boundary, we see that the product

$$s^{N-2}(P, Q)G(P, Q)$$

which is everywhere continuous in P and Q, has an upper bound K on the domain D. Thus $|c(Q)G(P, Q)| \leqslant Ks^{-N+2}$ where K is independent of P and Q. The estimates of § 6.4 can therefore be applied to the iterated kernels, and we conclude that these iterated kernels are continuous from a certain order on. Thus the Fredholm theorems apply to (8.4.5).

According to the third 'orthogonality' theorem, there exists a solution of (8.4.5) if and only if $v(P)$ is orthogonal on D to the solutions $w_\mu(P)$, if any, of the homogeneous transposed equation

$$w(P)-c(P) \int_D G(P, Q)w(Q) \, dV_Q = 0. \qquad (8.4.8)$$

As in the preceding section, we find that every solution $w_\mu(P)$ has the form

$$w_\mu(P) = c(P)z_\mu(P), \qquad (8.4.9)$$

where $z_\mu(P)$ is a solution of the homogeneous equation

$$z_\mu(P)- \int_D G(P, Q)c(Q)z_\mu(Q) \, dV_Q = 0. \qquad (8.4.10)$$

Clearly the $z_\mu(P)$ all vanish on the boundary, since the integral in (8.4.10) has this property. Also, applying the Laplacian to (8.4.10), we see that

$$\Delta z_\mu(P)+c(P)z_\mu(P) = 0. \qquad (8.4.11)$$

Thus the $z_\mu(P)$ are the eigenfunctions of the original differential equation on the domain D.

A solution of (8.4.5) exists, therefore, if and only if each of the integrals

$$\int_D v(P)w_\mu(P) \, dV_P$$

vanishes. With the help of (8.4.9), (8.4.11), and Green's formula, we see that the integral becomes

$$\int_D v(P)c(P)z_\mu(P) \, dV_P = - \int_D v(P)\Delta z_\mu(P) \, dV_P$$

$$= - \int_D z_\mu(P)\Delta v(P) \, dV_P - \int_B \left(v(p) \frac{\partial z_\mu(p)}{\partial n} - z_\mu(p) \frac{\partial v(p)}{\partial n} \right) dS_p. \qquad (8.4.12)$$

The last term disappears since $z_\mu(p) = 0$. Since $\Delta v(P) = \rho(P)$ and $v(p)$ may be replaced by $f(p)$, we see that the orthogonality condition becomes

$$\int_D \rho(P)z_\mu(P) \, dV_P + \int_B f(p) \frac{\partial z_\mu(p)}{\partial n} \, dS_p = 0. \qquad (8.4.13)$$

There appear here only the given data of the problem, and the eigenfunctions.

When (8.4.13) holds, a solution $u(P)$ of the integral equation exists. The verification that $u(P)$ is sufficiently regular to be a solution of the differential equation follows as in the preceding section. We conclude that *there exists a solution of* $\Delta u + cu = \rho$ *with boundary values* $f(p)$ *if and only if* (8.4.13) *holds for every eigenfunction* $z_\mu(P)$.

Thus, if the assigned boundary values are zero, the condition (8.4.13) is

$$\int_D \rho(P) z_\mu(P) \, dV_P = 0; \qquad (8.4.14)$$

it has the same form as the condition (8.3.12) in the case of a closed space. On the other hand, if the volume density $\rho(P)$ is zero, the condition (8.4.13) takes the form

$$\int_B f(p) \frac{\partial z_\mu(p)}{\partial n} \, dS_p = 0. \qquad (8.4.15)$$

That is, the Dirichlet problem for $\Delta u + cu = 0$ with boundary values $u(p) = f(p)$ on B, has a solution if and only if (8.4.15) holds for every eigenfunction $z_\mu(P)$.

Exercise 1. Show that equation (8.4.5) can be written as an equation with a symmetric kernel, if we take as unknown function

$$y(Q) = |c(Q)|^{\frac{1}{2}} u(Q).$$

Exercise 2. Show that a necessary condition for the existence of a solution of $\Delta u + cu = \rho$ on D, with $\partial u / \partial n = g$ on B, is that

$$\int_D z(P) \rho(P) \, dV_P - \int_B z(p) g(p) \, dS_p = 0$$

for every solution $z(P)$ of $\Delta u + cu = 0$ with $\partial z / \partial n = 0$.

Exercise 3. Find a necessary condition analogous to (8.4.13), in order that the equation

$$\Delta u + \mathbf{b} \cdot \nabla u + cu = \rho \quad \text{(on } D\text{)},$$

should possess a solution u with boundary values $u = f$ on B.

8.5. Eigenfunction expansions. The domain functionals, such as $G(P, Q)$ or $N(P, Q)$, associated with self-adjoint differential equations $\Delta u - qu = 0$, have played a double role in our previous work. In the first place they have been used to express the solutions of boundary value problems in terms of the assigned data. Then also, they have appeared as the kernels of integral equations the solution of which has led to further boundary value theorems, and other results concerning eigenfunctions. Since the Green's function of a self-adjoint differential equation is symmetric, we can apply to these integral equations the theory of symmetric kernels. In this way we shall find bilinear series expansions for these domain functionals.

For simplicity we shall restrict our treatment to the equation

$$\Delta u + \lambda u = 0 \quad (\lambda = \text{constant}) \tag{8.5.1}$$

on a given domain D. A typical eigenvalue problem for (8.5.1) is that of finding the eigenvalues λ to which correspond eigenfunctions vanishing on the boundary B. Let $G(P, Q)$ be the Green's function of Laplace's equation for D. Then solutions of (8.5.1) which vanish on D, i.e. the eigenfunctions, must satisfy the equivalent homogeneous integral equation

$$u(P) = \lambda \int_D G(P, Q) u(Q) \, dV_Q. \tag{8.5.2}$$

The equivalence of (8.5.1), together with the homogeneous boundary condition (zero data), and (8.5.2) becomes evident if we operate on (8.5.2) with Δ. Thus the eigenfunctions of the Dirichlet problem for (8.5.1) are the eigenfunctions of (8.5.2).

The kernel $G(P, Q)$ is symmetric, and has continuous iterated kernels. Thus, from § 6.5 we see that (8.5.2) has at least one eigenvalue λ_1 and a corresponding eigenfunction $u_1(P)$. Suppose that only a finite number of eigenvalues and eigenfunctions occur. Then, by (6.5.7), we see that

$$G(P, Q) = \sum \frac{u_n(P) u_n(Q)}{\lambda_n}, \tag{8.5.3}$$

where the $u_n(P)$ are the orthonormal eigenfunctions and the eigenvalues. If in (8.5.3) the series on the right terminates, then questions of convergence do not arise and the formula is valid for all P and Q, the right-hand side being everywhere continuous. But if $P \to Q$, we have $G(P, Q) \to \infty$ and the formula would not be correct. It follows, then, that the number of eigenvalues λ_n (each of which has a finite number of independent eigenfunctions) must be infinite. We shall assume that the infinite bilinear series (8.5.3) converges for $P \neq Q$; with this understanding it follows from § 6.5 that (8.5.3) is a valid equation.

The symmetry of $G(P, Q)$, which is characteristic of self-adjoint equations (8.5.1), shows that the eigenfunctions have the orthogonality property and that the eigenvalues are real. These facts can be proved directly from the differential equation and the boundary condition; the proofs we suggest as an exercise. In this special case we can deduce from the differential equation a further limitation, namely, that the eigenvalues are all positive. Indeed, an eigenfunction is a solution which vanishes on B and so must have in D either a positive maximum or a negative minimum. This contradicts the maximum principle (§ 5.2) if λ is negative. The value $\lambda = 0$ is not an eigenvalue, by § 5.1; hence $\lambda_n > 0$ as stated.

The iterated kernels have expansions

$$G^{(r)}(P, Q) = \sum_{n=1}^{\infty} \frac{u_n(P)u_n(Q)}{\lambda_n^r}, \qquad (8.5.4)$$

which converge uniformly if r is large enough. Thus the resolvent kernel $G(P, Q, \lambda)$ corresponding to $G(P, Q)$ has, according to (6.5.4), the formal expansion

$$G(P, Q, \lambda) = \sum_{r=0}^{\infty} G^{(r+1)}(P, Q)\lambda^r = \sum_{r=0}^{\infty} \lambda^r \sum_{n=1}^{\infty} \frac{u_n(P)u_n(Q)}{\lambda_n^{r+1}}$$

$$= \sum_{n=1}^{\infty} \frac{u_n(P)u_n(Q)}{\lambda_n} \sum_{r=0}^{\infty} \left(\frac{\lambda}{\lambda_n}\right)^r$$

$$= \sum_{n=1}^{\infty} \frac{u_n(P)u_n(Q)}{\lambda_n - \lambda}. \qquad (8.5.5)$$

Provided that λ is not equal to any eigenvalue λ_n, this series converges under the same conditions as does (8.5.3). According to this formula, $G(P, Q, \lambda)$ is the Green's function of that eigenvalue problem which has eigenfunctions $u_n(P)$ and eigenvalues $\lambda_n - \lambda$. That is, $G(P, Q, \lambda)$ is the Green's function for the equation $\Delta u + \lambda u = 0$. Or, equivalently, $G(P, Q, \lambda)$ is the kernel of the integral operator inverse to the differential operator $\Delta + \lambda$, with the boundary condition $u(p) = 0$. We now see directly that if λ is equal to an eigenvalue λ_n this Green's function is undefined because of the singularity in the corresponding term of the series in (8.5.5).

For example, let us consider the square $0 \leqslant x \leqslant \pi$, $0 \leqslant y \leqslant \pi$, in the Euclidean plane, and the equation

$$\frac{\partial^2 u}{\partial x^2} + \frac{\partial^2 u}{\partial y^2} + \lambda u = 0.$$

This equation has solutions of the form

$$\sin(mx - \delta_m)\sin(ny - \epsilon_n)$$

if $\lambda = m^2 + n^2$. The condition of vanishing when $x = 0$ requires $\delta_m = 0$; and when $x = \pi$, that m be an integer. Similarly, $\epsilon_n = 0$ and n must be integral if $u(P)$ is to be an eigenfunction of the Dirichlet problem. The normalization is easily found while the orthogonality of the various eigenfunctions is evident from the theory of Fourier sine series. The orthonormal eigenfunctions are

$$u_{mn}(P) = \frac{2}{\pi} \sin mx \sin ny,$$

and the Green's function is therefore

$$G(P,Q,\lambda) = \frac{4}{\pi^2} \sum_{m,n=1}^{\infty} \frac{\sin mx \sin ny \sin m\xi \sin n\eta}{m^2+n^2-\lambda}.$$

This series is known to converge when $(x-\xi)^2+(y-\eta)^2$ is positive, and λ is not an eigenvalue.

Exercise. Let h be a non-negative constant. Show that the eigenfunctions of the square $0 \leqslant x \leqslant \pi$, $0 \leqslant y \leqslant \pi$, for the above differential equation, and the boundary condition $(\partial u/\partial n)+hu = 0$, are

$$u_{mn} = \frac{2}{\pi} \sin(mx+\delta_m)\sin(ny+\delta_n),$$

where m, n are integers and $\tan \delta_m = -(m/h)$. Deduce formal expansions of the form (8.5.5) for the Robin and Neumann resolvent functions.

As another example, consider the sphere of radius a in Euclidean three-space. In spherical polar coordinates $x = r\sin\theta\cos\varphi$, $y = r\sin\theta\sin\varphi$, $z = r\cos\theta$, the differential equation $\Delta u+k^2u = 0$ has the form

$$\frac{\partial^2 u}{\partial r^2}+\frac{2}{r}\frac{\partial u}{\partial r}+\frac{1}{r^2\sin\theta}\frac{\partial}{\partial\theta}\left(\sin\theta\frac{\partial u}{\partial\theta}\right)+\frac{1}{r^2\sin^2\theta}\frac{\partial^2 u}{\partial\varphi^2}+k^2u = 0,$$

where we have written k^2 in place of λ.

Separation of the variables shows that

$$j_n(kr)P_n^m(\cos\theta)e^{im\varphi} \quad (m,n \text{ integers}, -n \leqslant m \leqslant n)$$

are solution functions single-valued in the space. Here $j_n(x)$ denotes the spherical Bessel function $x^{-\frac{1}{2}}J_{n+\frac{1}{2}}(x)$, and k is undetermined. These functions vanish for $r = a$ provided that

$$j_n(ka) = 0;$$

let k_{np} $(p = 1, 2,...)$, denote the roots of this equation. The eigenfunctions of the Dirichlet problem are then

$$u_{mnp} = j_n(k_{np}r)P_n^m(\cos\theta)e^{im\varphi}.$$

From the general theorem on orthogonality, we see that these eigenfunctions are mutually orthogonal if they have different eigenvalues k_{np}^2. This is assured if the values of n are equal and those of p distinct. For distinct values of the index n, and equal values of m, the factors $P_n^m(\cos\theta)$ are orthogonal. Finally, if the values of m differ, the factors $e^{im\varphi}$ lead to orthogonality. Hence the integral over a product $u_{mnp}u_{m'n'p'}$ is different from zero only if $n = n'$, $m = m'$, $p = p'$. Let us denote the normalized eigenfunctions by

$$u_{mnp} = \psi_n(k_{np}\, r)\, \Pi_n^m(\cos \theta)e^{im\varphi}.$$

The differential equation is invariant under any rotation of the sphere, and we are free to choose any point for the 'north pole' of the coordinate system and any meridian plane for the plane $\varphi = 0$. In forming the Green's function, we might then expect that it can be expressed in terms of r_P and r_Q, and the angle Θ subtended at the origin by the two points P and Q. We have

$$\cos \Theta = \cos \theta_P \cos \theta_Q + \sin \theta_P \sin \theta_Q \cos(\varphi_P - \varphi_Q).$$

This expectation is realized, because the normalized associated Legendre functions satisfy the addition theorem

$$(n+\tfrac{1}{2})^{\frac{1}{2}}\, \Pi_n^0(\cos \Theta) = \sum_{m=-n}^{n} \Pi_n^m(\cos \theta_P)\, \Pi_n^m(\cos \theta_Q)e^{im(\varphi_P - \varphi_Q)},$$

which expresses the surface harmonic $\Pi_n^0(\cos \Theta)$, in the coordinate system wherein $\theta_Q = 0$, $\varphi_P = 0$, as a sum of surface harmonics in the original coordinate system. For a proof of this addition theorem, we refer the reader to (36, p. 328).

The representation (8.5.5) has in our present case the form

$$G(P, Q, \lambda) = \sum_{p=0}^{\infty} \sum_{n=0}^{\infty} \frac{\psi_n(k_{np}\, r_P)\psi(k_{np}\, r_Q)}{k_{np}^2 - \lambda}\, (n+\tfrac{1}{2})^{\frac{1}{2}}\, \Pi_n^0(\cos \Theta).$$

In this expansion there appear only r_P, r_Q and the angle Θ.

To sum up, we see that the bilinear series expansion provides a quite general method for finding the Green's function or other similar domain functionals, provided the eigenvalues and eigenfunctions are known. However, the convergence is often hard to discuss, and explicit summation of the bilinear series is practicable only in the simplest cases.

The bilinear formula also leads in a formal sense to the expansion of an arbitrary function in a series of eigenfunctions. Let $f(P)$ be a function which vanishes on B. Then, from (7.3.8) we see that

$$f(P) = - \int_D G(P, Q)\Delta f(Q)\, dV_Q.$$

Replacing $G(P, Q)$ by its series expansion, and inverting the order of summation and integration, we have, formally,

$$f(P) = - \int_D \sum \frac{u_n(P)u_n(Q)}{\lambda_n} \Delta f(Q)\, dV_Q = \sum_n f_n\, u_n(P),$$

where the expression for the coefficient f_n is easily written down. The validity of the Fourier expansion may thus be verified if the inversion of summation and integration can be justified. We remark that if $f(P)$ is sufficiently differentiable, a representation of the above form by means of continuous iterated kernels is possible, and the inversion is then permissible. Since the Fourier coefficients of any function with respect to a given orthogonal set such as $u_n(P)$ are found in (6.6.1), the values indicated above for f_n must agree with those in (6.6.1). The reader will easily verify that this is so with the help of Green's formula.

Exercise 1. Find the eigenfunctions of the two-dimensional spherical surface with metric (8.3.21). Write down the series expansion for the resolvent and for the Green's function (8.3.22). *Hint*: $\lambda_n = n(n+1)$ is a $(2n+1)$-fold eigenvalue.

Exercise 2. Show that the equation

$$\Delta u - q(P)u = \lambda \rho(P)u,$$

where $q(P) > 0$, $\rho(P) \neq 0$, has infinitely many eigenvalues and eigenfunctions for the Dirichlet problem on a given domain.

8.6. Initial value problems. The application of eigenfunction expansions to initial value problems for parabolic or hyperbolic equations is best seen by means of examples. Consider the problem of determining the temperature distribution in a domain D, when an initial temperature $T(P)$ is prescribed, and when the temperature on the boundary B is held fixed at the value $T(p)$. The differential equation we shall take to be

$$\frac{\partial u}{\partial t} = \Delta u \quad (u = u(P,t)), \tag{8.6.1}$$

where t is the time, and u the temperature to be determined. The dependence upon the time can, under these assumptions, be separated out. In fact, the product

$$e^{-\lambda t} u(P) \tag{8.6.2}$$

satisfies (8.6.1) provided that $\Delta u(P) + \lambda u(P) = 0$.

To construct a solution of the given problem we shall first 'subtract out' the corresponding equilibrium problem—that is, we shall reduce the assigned initial distribution on the boundary to zero by subtracting a solution function which is independent of t and has boundary values $T(p)$. Such a function exists, being in fact that harmonic function $v(P)$ with boundary values equal to $T(p)$.

Now define $\qquad\qquad f(P) = T(P) - v(P);$ $\qquad\qquad$ (8.6.3)

we see that $f(p)$ is zero. Let $f(P)$ be expanded in a series of orthonormal eigenfunctions $u_n(P)$ of $\Delta u + \lambda u = 0$, which vanish on the boundary:

$$f(P) \sim \Sigma f_n u_n(P), \qquad f_n = \int_D f(P) u_n(P) \, dV_P. \qquad (8.6.4)$$

We now see that the solution of the problem is given, formally at least, by the series

$$u(P,t) = v(P) + \sum_{n=1}^{\infty} f_n e^{-\lambda_n t} u_n(P). \qquad (8.6.5)$$

Formal verification that this function has the requisite properties is immediate. Since all of the λ_n are positive, the terms in the sum decrease as t increases. Indeed, the limit of these terms, as t tends toward infinity, is zero, and the temperature tends to the equilibrium distribution $v(P)$ as $t \to \infty$. Thus the solution of the parabolic equation (8.6.1) tends in the limit to a harmonic function, the solution of an elliptic equation. We remark that the series (8.6.5) may converge for positive values of t even if the Fourier series for $f(P)$ is not convergent, since the exponentials act as convergence factors. In particular cases of this kind it is often possible to show that the limit of $u(P,t)$ as $t \to 0$ is equal to $f(P)$, at least where $f(P)$ is continuous.

When the boundary temperature is fixed at zero, the harmonic function $v(P)$ drops out of the solution. Since $f_n = \int_D f(Q) u_n(Q) \, dV_Q$ we see that (8.6.5) has now the form

$$u(P,t) = \int_D \bar{G}(P,Q,t) f(Q) \, dV_Q, \qquad (8.6.6)$$

where $\qquad\qquad \bar{G}(P,Q,t) = \sum_{n=1}^{\infty} e^{-\lambda_n t} u_n(P) u_n(Q).$ $\qquad$ (8.6.7)

Thus $G(P,Q,t)$ plays the role of a Green's function for the parabolic differential equation. The physically evident fact that a positive initial distribution of temperature remains positive corresponds to a property of the Green's function (which we shall not attempt to prove here), namely,

$$\bar{G}(P,Q,t) \geqslant 0. \qquad (8.6.8)$$

A formal calculation with the orthonormal functions $u_n(P)$ shows that

$$\int_D \bar{G}(P,R,t_1) \bar{G}(R,Q,t_2) \, dV_Q = \bar{G}(P,Q,t_1+t_2). \qquad (8.6.9)$$

This relation shows that we will find the same distribution at time t_1+t_2 whether we calculate from the initial time $t = 0$ directly by means of

(8.6.6), or in two stages using the time t_1 as an intermediate step. Thus (8.6.9) is an equation of consistency for the diffusion process described by the parabolic equation.

A somewhat different problem is suggested by the temperature distribution in a region D insulated so that no heat passes across the boundary; this corresponds to the Neumann boundary condition

$$\frac{\partial u(p)}{\partial n} = 0. \tag{8.6.10}$$

If now $u(P, 0) = f(P)$, where the normal derivative of $f(P)$ on B is zero, we can expand $f(P)$ in a series $\sum f_n u_n(P)$ of eigenfunctions of the problem

$$\Delta u + \lambda u = 0, \qquad \frac{\partial u(p)}{\partial n} = 0. \tag{8.6.11}$$

The lowest eigenvalue is $\lambda_0 = 0$, the corresponding eigenfunction is a constant. Thus the coefficient f_0 is equal to the average value of $f(P)$ over D, times $\sqrt{V}$ (V volume of D). Again the formal solution of the problem is

$$u(P, t) = \sum_{n=0}^{\infty} e^{-\lambda_n t} f_n u_n(P). \tag{8.6.12}$$

As $t \to \infty$, all terms save the first decrease exponentially. The limit of $u(P, t)$ as $t \to \infty$ is thus the average value

$$f_0 u_0(P) = V^{-\frac{1}{2}} \int_D f(P) V^{-\frac{1}{2}} \, dV_P$$

of the given initial distribution $f(P)$.

The hyperbolic equation

$$\frac{\partial^2 u}{\partial t^2} = \Delta u \quad (u = u(P, t)), \tag{8.6.13}$$

where again Δ is the Laplace operator based on a positive definite Riemannian metric, may also be separated in this way. The solutions are seen to be

$$\cos kt . u(P), \qquad \sin kt . u(P), \tag{8.6.14}$$

where again $\qquad \Delta u + k^2 u = 0 \quad (\lambda = k^2). \tag{8.6.15}$

Solutions of the cosine type have non-vanishing values but vanishing time-derivative when $t = 0$, the reverse being true for the sine solutions.

Applications such as the vibrations of membranes or electromagnetic oscillations lead to the following type of problem. Let the boundary condition be

$$u(p, t) = 0 \quad (p \text{ on } B) \tag{8.6.16}$$

and the initial conditions

$$u(P, 0) = f(P) \quad (f(p) = 0),$$
$$u_t(P, 0) = g(P) \quad (g(p) = 0). \tag{8.6.17}$$

The formal solution follows immediately upon an expansion of $f(P)$ and $g(P)$ in a series of eigenfunctions of (8.6.15) which vanish on the boundary.

Let
$$f(P) = \sum_{n=1}^{\infty} f_n u_n(P), \qquad g_n(P) = \sum_{n=1}^{\infty} g_n u_n(P). \tag{8.6.18}$$

Then

$$u(P, t) = \sum_{n=1}^{\infty} f_n \cos(k_n t) u_n(P) + \sum_{n=1}^{\infty} \frac{g_n}{k_n} \sin(k_n t) u_n(P) \tag{8.6.19}$$

is easily seen to be the formal solution of the problem. The convergence or summability of this series is a matter for study in particular cases. The time-dependent factors oscillate with undiminished amplitude, in contrast to the parabolic case.

For the success of this method, the rather special nature of the initial and boundary conditions is necessary, as well as the separable character of the differential equation. However, these restrictions are satisfied in a great variety of practical applications.

Exercise 1. For the equation $u_t = u_{xx}$, $-\infty < x < \infty$, $t > 0$, the solutions of the form (8.6.2) are $(\lambda = k^2)$

$$\frac{1}{\sqrt{(2\pi)}} e^{-k^2 t} \begin{cases} \cos kx \\ \sin kx \end{cases} \quad (-\infty < k < \infty).$$

Show that the Green's function (8.6.7) is

$$\frac{1}{\sqrt{(4\pi t)}} \exp\left[-\frac{(x-\xi)^2}{4t} \right].$$

Exercise 2. For the Green's function (8.6.2), the resolvent (8.5.5), and the iterated kernels (8.5.4), the following formal relations are indicated:

$$\int_0^{\infty} e^{\lambda t} \bar{G}(P, Q, t) \, dt = G(P, Q, \lambda),$$

$$\int_0^{\infty} t^{p-1} \bar{G}(P, Q, t) \, dt = \Gamma(p) G^{(p)}(P, Q).$$

Exercise 3. In Euclidean N-space, show that spherically symmetric solutions of (8.6.1) satisfy

$$\frac{\partial u}{\partial t} = \frac{\partial^2 u}{\partial r^2} + \frac{N-1}{r} \frac{\partial u}{\partial r}.$$

Show that this equation has solutions

$$e^{-k^2 t} r^{-\frac{1}{2}(N-2)} J_{\frac{1}{2}(N-2)}(ikr),$$

$$t^{-\frac{1}{2}N} \exp[-r^2/4t].$$

Exercise 4. Discuss the initial value problem for Schrödinger's equation $(i^2 = -1)$

$$i\hbar \frac{\partial \psi}{\partial t} = -\frac{\hbar^2}{2m} \Delta\psi + V(P)\psi.$$

(Here the separation constant λ is E the energy of the state represented by the eigenfunction.)

Exercise 5. On a closed space F, find the solution of $u_t = \Delta u - q(P)u$, $q(P) \geqslant 0$, $u(P, 0) = f(P)$; and calculate the limit of $u(P, t)$ as $t \to \infty$ in the two cases (1) $q(P) \not\equiv 0$, (2) $q(P) \equiv 0$.

Exercise 6. If $R^2 = r^2 + r_0^2 - 2rr_0 \cos\theta$, and if the generalized Legendre coefficients $P_n(\cos\theta|p)$ are defined by the expansion

$$\frac{1}{R^p} = \frac{1}{r_0^p} \sum_{n=0}^{\infty} \left(\frac{r}{r_0}\right)^n P_n(\cos\theta|p) \quad (r < r_0),$$

show that

$$\frac{d}{d\theta}\left(\sin^p\theta \, \frac{dP_n(\cos\theta|p)}{d\theta}\right) + n(n+p)\sin^p\theta P_n(\cos\theta|p) = 0,$$

and that $\displaystyle\int_0^\pi P_n(\cos\theta|p)P_m(\cos\theta|p)\sin^p\theta \, d\theta = 0 \quad (m \neq n).$

Exercise 7. Show that $r^n P_n(\cos\theta|p)$ and $r^{-n-p}P_n(\cos\theta|p)$ are harmonic functions in Euclidean space of $N = p+2$ dimensions. Also show that

$$r^{-\frac{1}{2}p} Z_{n+\frac{1}{2}p}(kr)P_n(\cos\theta|p)$$

is a solution of $\Delta u + k^2 u = 0$, where $Z_{n+\frac{1}{2}p}(x)$ is any solution of Bessel's equation of order $n + \frac{1}{2}p$.

IX

NORMAL HYPERBOLIC EQUATIONS

THE classification made in Chapter IV of linear partial differential equations of the second order was based upon the properties of the quadratic form

$$Q(\xi) = a^{ik}\xi_i\xi_k,$$

where the a^{ik} are coefficients of the second derivatives in the differential equation. We shall now study normal hyperbolic equations, or wave equations, for which $Q(\xi)$, when referred to its principal axes, contains one term differing in sign from the remainder. The signature may be taken as $(1, N-1)$, there being one positive and $N-1$ negative coefficients. In physical applications, these correspond to one time-like and $N-1$ space-like dimensions. The vanishing of the indefinite quadratic form $Q(\xi)$ singles out certain characteristic surface elements ξ_i, which take a leading role in the geometric theory.

The solution functions may be interpreted as waves, the variation with time at a fixed point in space corresponding to the undulation of the wave. The discontinuities and singularities, and the manner of propagation of the waves through space, are closely related to the characteristic surfaces. So also is the formulation of the Cauchy initial value problem largely determined by the nature of these surfaces. In this chapter we study these aspects of the theory of the characteristics, in preparation for the integration of the wave equation which is carried out in Chapter X.

9.1. Characteristic surfaces. Given the differential equation

$$L[u] = \Delta u + \mathbf{b} \cdot \nabla u + cu$$

$$= a^{ik}\frac{\partial^2 u}{\partial x^i \partial x^k} + \beta^i \frac{\partial u}{\partial x^i} + cu = 0 \tag{9.1.1}$$

of normal hyperbolic type, we may formulate the following initial value problem. On a surface S: $\varphi(x^i) = 0$ let values of the unknown function $u(x^i)$ and of its first partial derivatives $p_i(x^j)$ be assigned, these values satisfying the 'strip condition'

$$du = p_i\,dx^i \tag{9.1.2}$$

for all displacements tangent to the surface S. We may imagine that S is parametrized by $N-1$ parameters $\lambda^1,..., \lambda^{N-1}$; $x^i = x^i(\lambda^\alpha)$, so that we are given the $N+1$ functions $u(\lambda^\alpha)$, $p_i(\lambda^\alpha)$ $(i = 1,...,N)$. These satisfy, corresponding to (9.1.2), the conditions

$$\frac{\partial u}{\partial \lambda^\alpha} = p_i \frac{\partial x^i}{\partial \lambda^\alpha} \quad (\alpha = 1,...,N-1). \tag{9.1.3}$$

We require to find a solution of (9.1.1) defined in a region containing S, which on S agrees in value with u and whose first partial derivatives agree with the assigned functions p_i. Such a solution would be said to contain the given first order strip consisting of the functions

$$x^i(\lambda^\alpha), \qquad u(\lambda^\alpha), \qquad p_i(\lambda^\alpha).$$

Our problem is, then, to enlarge this first order strip to a portion of an integral surface of (9.1.1).

For this purpose it is convenient, even for non-analytic equations, to adopt the point of view of § 1.3, and to inquire whether the higher order derivatives of a solution function may be determined in a unique way from the given data. We therefore attempt to calculate, from the data and the differential equation, the values on S of the higher derivatives of u. Let us first choose a coordinate system in which the surface S becomes a hyperplane. Such a system of coordinates is given by the variables λ^α $(\alpha = 1,...,N-1)$ and the further variable $\lambda^N = \varphi(x^i)$, which is zero on S. In § 4.1 it was seen that the coefficients a^{ik} of the second derivatives transform as components of a contravariant tensor of the second order:

$$a'^{rs} = a^{ik} \frac{\partial x'^r}{\partial x^i} \frac{\partial x'^s}{\partial x^k} \quad (x'^\alpha \equiv \lambda^\alpha, \, x'^N \equiv \varphi), \tag{9.1.4}$$

so that in particular

$$a'^{NN} = a^{ik} \frac{\partial \varphi}{\partial x^i} \frac{\partial \varphi}{\partial x^k} = Q(\varphi). \tag{9.1.5}$$

On the surface $S : x'^N = 0$, we are given u, and the N first partials $p_i = \partial u/\partial x^i$, as functions of the $N-1$ variables $x'^\alpha = \lambda^\alpha$ $(\alpha = 1,...,N-1)$. Thus we can find all derivatives of these with respect to the λ^α variables. Now, writing (9.1.1) in the new coordinate system, we see that, to determine the second derivative $\dfrac{\partial^2 u}{\partial \varphi^2} = \dfrac{\partial^2 u}{(\partial x'^N)^2}$, we have

$$L[u] = a'^{rs} \frac{\partial^2 u}{\partial x'^r \partial x'^s} + \beta'^r \frac{\partial u}{\partial x'^r} + cu$$

$$= a'^{NN} \frac{\partial^2 u}{(\partial x'^N)^2} + ... + cu \tag{9.1.6}$$

$$= Q(\varphi) \frac{\partial^2 u}{\partial \varphi^2} + ... + cu = 0.$$

The terms not indicated explicitly are all given as functions of the $x'^\alpha = \lambda^\alpha$ on S.

If $Q(\varphi)$ is different from zero, the second derivative $\partial^2 u/\partial\varphi^2$ is uniquely determined on S. Similarly, if we differentiate the equation (9.1.1), we find relations from which all derivatives of u with respect to φ may be determined in succession, provided that $Q(\varphi)$ is not zero. If now the differential equation is analytic, we might use the dominating power series technique to show that a solution of the problem exists. In this case, it is clear that the solution is well-determined. Provided that $Q(\varphi) \neq 0$, i.e. that the surface $\varphi = 0$ is nowhere characteristic, an analytic solution of the Cauchy initial value problem is unique.

Consider now the exceptional case when $Q(\varphi)$ is zero, and the higher derivatives cannot be so determined. For simplicity, we shall assume that $Q(\varphi)$ vanishes identically on S:

$$Q(\varphi) = a^{ik} \frac{\partial\varphi}{\partial x^i} \frac{\partial\varphi}{\partial x^k} \equiv 0. \tag{9.1.7}$$

Since $\varphi = \varphi(x^i)$ depends only on the x^i, as do the a^{ik}, the vanishing of $Q(\varphi)$ is determined by the surface S alone and not in any way by the data u, p_i assigned on S. The surface S is now said to be a *characteristic surface*. Since $Q(\varphi)$ is an invariant, the fact that S is characteristic is independent of any coordinate system.

If S is characteristic, we may observe two facts; first, the lack of determination of the solution, upon which we have already remarked; second, the nature of the differential equation on S. Since $Q(\varphi)$ is zero, the only term in $L(u)$ not determined by the given data drops out of the equation. The statement $L(u) = 0$ must then constitute a necessary condition on the data of the problem. No solution will be possible unless this condition of consistency is fulfilled. In the next section we shall pursue this question in more detail.

These facts may be presented in a somewhat different light, which leads to another point of view regarding characteristic surfaces. On a characteristic surface S, the differential operator $L[u]$ contains derivatives of the first and second orders in directions tangential to S, and also derivatives of the first order, and mixed derivatives, transverse to S; but it contains no second derivative $u_{\varphi\varphi}$ across or transverse to S, since the corresponding coefficient $a'^{NN} = Q(\varphi)$ is zero. Hence $L[u]$ is an inner differential operator for the first order strip, that is, $L[u]$ contains only quantities which can be found from the data of the strip. Thus $L[u]$ is an inner differential operator for all first order strips based on a characteristic surface, and for such strips only.

The condition $Q(\varphi) = 0$ which determines characteristic surfaces has the form of a partial differential equation of the first order for the function $\varphi(x^i)$. However, a surface $\varphi(x^i) = 0$ may be characteristic even if this equation is not satisfied *identically* in the variables x^i. Indeed, $\varphi = 0$ is characteristic if $Q(\varphi)$ is zero as a consequence of $\varphi = 0$; that is, if $Q(\varphi)$ is zero on S itself, but not identically zero. If $Q(\varphi)$ is zero identically, then the family of surfaces $\varphi(x^i) = c$ are all characteristic, since $Q(\varphi - c) \equiv Q(\varphi)$, and the surface $\varphi = 0$ is then a member of a family of characteristic surfaces which fills a certain region of the space. We now wish to show that any characteristic surface S is in fact a member of such a family.

If the surface S is given by an equation of the form

$$x^N = \psi(x^1, ..., x^{N-1}),$$ (9.1.8)

we may express the characteristic condition (9.1.7) by taking

$$\varphi = \psi(x^1, ..., x^{N-1}) - x^N.$$

Thus we find

$$a^{\alpha\beta} \frac{\partial \psi}{\partial x^\alpha} \frac{\partial \psi}{\partial x^\beta} - 2a^{\alpha N} \frac{\partial \psi}{\partial x^\alpha} + a^{NN} = 0$$ (9.1.9)

as the form of the condition, summation over α, β being understood. This equation we may interpret as a partial differential equation of first order for $\psi(x^\alpha)$. Any solution $\psi(x^\alpha)$ of (9.1.9) corresponds to a surface (9.1.8) which is certainly characteristic. Now, let

$$\psi(x^1, ..., x^{N-1}, c)$$

be a one-parameter family of solutions of (9.1.9), such that

$$\psi(x^1, ..., x^{N-1}, 0) = \psi(x^1, ..., x^{N-1}).$$

Then the family of surfaces

$$x^N = \psi(x^1, ..., x^{N-1}, c)$$ (9.1.10)

are all characteristic, and moreover, (9.1.8) is a member of the family. Solving (9.1.10) for c, we find the equation of the family in the form $\bar{\varphi}(x^1, ..., x^N) = c$, where $Q(\bar{\varphi}) = 0$, in view of (9.1.9).

By way of example, let us take the wave-equation

$$u_{tt} - u_{xx} - u_{yy} - u_{zz} = 0.$$

The condition for characteristic surfaces $\varphi = c$ is

$$\varphi_t^2 - \varphi_x^2 - \varphi_y^2 - \varphi_z^2 = 0.$$

Considered as a partial differential equation of the first order, this equation has a complete integral of the form

$$\varphi = \alpha_0 t + \alpha_1 x + \alpha_2 y + \alpha_3 z + \beta,$$

where $\qquad\qquad \alpha_0^2 = \alpha_1^2 + \alpha_2^2 + \alpha_3^2.$

The planes represented by this complete integral depend on four parameters, say α_1, α_2, α_3, and β. These planes are all characteristic surfaces. So also must be the envelope of the family of these planes which pass through a fixed point, say the origin. This envelope is the cone

$$\varphi_1 \equiv t^2 - x^2 - y^2 - z^2 = 0.$$

Now φ_1 satisfies the partial differential equation

$$Q(\varphi_1) = \varphi_{1t}^2 - \varphi_{1x}^2 - \varphi_{1y}^2 - \varphi_{1z}^2 = 4\varphi_1,$$

so that the surface $\varphi_1 = c \neq 0$ is not characteristic. However, the equation of the cone may be written

$$t = (x^2 + y^2 + z^2)^{\frac{1}{2}},$$

and (9.1.9) becomes $\qquad \psi_x^2 + \psi_y^2 + \psi_z^2 = 1.$

This equation is satisfied by

$$\psi(x, y, z) = (x^2 + y^2 + z^2)^{\frac{1}{2}} + c,$$

so that the family of surfaces

$$\bar{\varphi} \equiv t - (x^2 + y^2 + z^2)^{\frac{1}{2}} = c$$

are all characteristic. The reader will verify that $\bar{\varphi}$ satisfies $Q(\bar{\varphi}) = 0$, identically.

Exercise 1. Find the planes and cones which are characteristic surfaces for

$$u_{tt} - u_{x^1 x^1} - \ldots - u_{x^M x^M} = 0.$$

Exercise 2. Show that the surfaces

$$\int \left(a_t^2 - \frac{a_\varphi^2}{\sin^2\theta} \right)^{\frac{1}{2}} d\theta + a_t t + a_\varphi \varphi = c$$

are characteristic surfaces of

$$\frac{\partial^2 u}{\partial t^2} = \frac{1}{\sin\theta} \frac{\partial}{\partial\theta} \left(\sin\theta \frac{\partial u}{\partial\theta} \right) + \frac{1}{\sin^2\theta} \frac{\partial^2 u}{\partial\varphi^2}.$$

Exercise 3. Show that $ds^2 = 0$ for the displacement normal to a characteristic surface, in the sense of § 4.2.

9.2. Bicharacteristics. The differential equation $L[u] = 0$ defines a Riemannian metric

$$ds^2 = a_{ik}\, dx^i dx^k \tag{9.2.1}$$

on the space of the independent variables x^i, as in § 4.2. The metric which corresponds to a hyperbolic equation is indefinite, its signature being that of the quadratic form $Q(\varphi)$, which is $(1, N-1)$ for a normal hyperbolic equation. Given a surface element, say $\partial\varphi/\partial x^i$, there is defined the displacement

$$dx^i = a^{ik} \frac{\partial\varphi}{\partial x^k}\, dt, \tag{9.2.2}$$

normal to the surface $\varphi = $ constant, in the Riemannian sense. In the theory of hyperbolic equations this direction has also been known as the co-normal or transversal. The normal derivative of a function ψ is given, apart from a constant factor, by

$$\frac{d\psi}{dn} = a^{ik} \frac{\partial\varphi}{\partial x^i} \frac{\partial\psi}{\partial x^k} = \nabla\varphi . \nabla\psi. \tag{9.2.3}$$

We now show that the normal to a characteristic surface S: $\varphi = c$ has a most striking property. *The normal is in fact tangential to the surface.* To see this we calculate the normal derivative of φ itself. We find

$$\frac{d\varphi}{dn} = a^{ik} \frac{\partial\varphi}{\partial x^i} \frac{\partial\varphi}{\partial x^k} = (\nabla\varphi)^2 = Q(\varphi) = 0, \tag{9.2.4}$$

since $\varphi = c$ is a characteristic surface. Thus φ does not change in numerical value under a displacement along the normal, so that the normal direction must be tangential to the surface $\varphi = c$ as stated. Conversely, if the normal has this property, we see that $d\varphi/dn$ is zero, and from (9.2.4) it follows that $\varphi = c$ must be a characteristic surface.

To a family of surfaces $\varphi(x^i) = c$, whether characteristic or not, corresponds a set of curves which, in the metric (9.2.1), are the orthogonal trajectories of the surfaces. These orthogonal curves are just the integral curves of the differential equations (9.2.2). But if the family of surfaces $\varphi(x^i) = c$ is characteristic, each orthogonal curve lies entirely in one of the surfaces, since, at each point, the orthogonal direction is tangent to the characteristic surface through that point. Another statement of the same result is that, since $(\nabla\varphi)^2 = 0$, the relation $\varphi(x^i) = $ constant is a first integral of the ordinary differential equations (9.2.2). Starting from any point P, we can construct an orthogonal trajectory which lies in the surface $\varphi(x^i) = c$, where $c = \varphi(P)$. Each characteristic surface, being of dimension $N-1$, may be regarded as being generated by an $(N-2)$-parameter family of the orthogonal curves.

This last geometrical remark suggests, as we shall now demonstrate analytically, that the orthogonal trajectories are precisely the character-istic curves of the first order characteristic equation

$$(\nabla\varphi)^2 = a^{ik} \frac{\partial\varphi}{\partial x^i} \frac{\partial\varphi}{\partial x^k} = 0. \qquad (9.2.5)$$

Setting
$$2F(x^i, p_k) = a^{ik} p_i p_k \quad \left(p_i \equiv \frac{\partial\varphi}{\partial x^i}\right), \qquad (9.2.6)$$

we see from (3.3.2) that the characteristic curves are integral curves of

$$\frac{dx^i}{dt} = \frac{\partial F}{\partial p_i}, \qquad \frac{dp_i}{dt} + \frac{\partial F}{\partial x^i} = 0. \qquad (9.2.7)$$

As was shown in § 3.1, the equations (9.2.7) have the first integral $F(x^i, p_k) = $ constant. For the characteristic curves we have $F(x^i, p_k) = 0$ —the constant is then zero. Since

$$\frac{\partial F}{\partial p_i} = a^{ik} p_k = a^{ik} \frac{\partial\varphi}{\partial x^k},$$

we see that the first of (9.2.7) is identical with (9.2.2). If, as we may suppose in constructing the orthogonal trajectories, the surfaces $\varphi(x^i) = c$ are given, then these curves are altogether determined by the equivalent relations (9.2.2) or (9.2.7 a). Therefore the orthogonal trajectories are identical with the characteristic curves of the characteristic equation. The name bicharacteristics has been given to these curves to emphasize this latter relationship. The term characteristic ray has also been used.

The bicharacteristics of the wave equation

$$u_{tt} - u_{xx} - u_{yy} - u_{zz} = 0$$

are the characteristic curves of

$$\varphi_t^2 - \varphi_x^2 - \varphi_y^2 - \varphi_z^2 \equiv p_t^2 - p_x^2 - p_y^2 - p_z^2 = 0.$$

Since the independent variables t, x, y, z do not appear explicitly, we see from (9.2.7 b) that

$$p_t = \alpha_0, \qquad p_x = \alpha_1, \qquad p_y = \alpha_2, \qquad p_z = \alpha_3,$$

where we must have $\alpha_0^2 = \alpha_1^2 + \alpha_2^2 + \alpha_3^2$, if the characteristic equation is to hold. Thus from (9.2.7 a) we find, since the p's are constants, that the characteristic rays (with parameter τ) are

$$t = t_0 + \alpha_0\tau, \qquad x = x_0 + \alpha_1\tau, \qquad y = y_0 + \alpha_2\tau, \qquad z = z_0 + \alpha_3\tau.$$

The rays passing through the point t_0, x_0, y_0, z_0, generate the characteristic cone

$$(t-t_0)^2 = (x-x_0)^2+(y-y_0)^2+(z-z_0)^2,$$

which is, as we already know, a characteristic surface.

Exercise. Find the parametric form of the bicharacteristics of the equation in Exercise 2, § 9.1. Find the equation of the cone generated by the bicharacteristics passing through the north pole $\theta = 0$ when $t = 0$.

The bicharacteristics have an interesting property with respect to the Riemannian metric (9.2.1), for they are the geodesic null lines of this metric. We may define the null lines by the following analogy with mechanics. Imagine a free particle of unit mass, moving along a curve C: $x^i = x^i(\sigma)$ (where σ is a parameter of position on the curve) whose kinetic energy T and Lagrange function L are given by

$$2L = 2T = a_{ik}\dot{x}^i\dot{x}^k = \left(\frac{ds}{d\sigma}\right)^2 \quad \left(\dot{x}^i \equiv \frac{dx^i}{d\sigma}\right). \tag{9.2.8}$$

(We have, in general, $L = T-V$ but here V is taken to be zero.) According to Hamilton's variational principle (21), the motion of the particle will be determined by a minimum (or extremum) of the integral

$$\int L(x^i, \dot{x}^i)\, d\sigma. \tag{9.2.9}$$

The extremum curves for this variational problem satisfy the Euler–Lagrange equations

$$\frac{d}{d\sigma}\left(\frac{\partial L}{\partial \dot{x}^i}\right) - \frac{\partial L}{\partial x^i} = 0. \tag{9.2.10}$$

Now $L(x^i, \dot{x}^i)$ is homogeneous of the second degree in the $\dot{x}^i$ variables. Making use of this, and of (9.2.10), we have

$$\frac{dL}{d\sigma} = \frac{\partial L}{\partial x^k}\dot{x}^k + \frac{\partial L}{\partial \dot{x}^k}\ddot{x}^k = \frac{d}{d\sigma}\left(\frac{\partial L}{\partial \dot{x}^k}\right)\dot{x}^k + \frac{\partial L}{\partial \dot{x}^k}\ddot{x}^k = \frac{d}{d\sigma}\left(\dot{x}^k \frac{\partial L}{\partial \dot{x}^k}\right) = 2\,\frac{dL}{d\sigma}.$$

Therefore $dL/d\sigma$ must vanish, so that L is a constant along the curve. This is the conservation of energy, in the mechanical analogy.

If L is a non-zero constant, we may divide (9.2.10) by $|L|^{\frac{1}{2}}$ and thus obtain

$$\frac{d}{d\sigma}\left(\frac{\partial |L|^{\frac{1}{2}}}{\partial \dot{x}^i}\right) - \frac{\partial |L|^{\frac{1}{2}}}{\partial x^i} = 0.$$

These are the Euler equations of the integral

$$\int |L|^{\frac{1}{2}}\, d\sigma = \int |a_{ik}\, dx^i dx^k|^{\frac{1}{2}} = \int |ds|.$$

That is, the path has extremal length in the Riemann metric and so is a geodesic in the more usual sense. If L has the value zero, the curve is known as a geodesic null line, since, in view of (9.2.8), $ds = 0$ all along the curve.

In order to show that the geodesic null lines are the bicharacteristics, we must transform the Lagrangian differential equations (9.2.10) into the Hamiltonian form (9.2.7). We define the momentum variables p_i by the relations

$$p_i = \frac{\partial L}{\partial \dot{x}^i} = a_{ik} \dot{x}^k; \qquad \dot{x}^k = a^{ik} p_i. \tag{9.2.11}$$

Thus the momenta are the covariant components of the velocity vector $\dot{x}^i$. Eliminating the $\dot{x}^i$ from the Lagrangian by means of (9.2.11) we obtain the Hamiltonian

$$H(x^i, p_k) = \tfrac{1}{2} a^{ik} p_i p_k = \tfrac{1}{2} p^2 = L(x^i, x^j). \tag{9.2.12}$$

The Hamiltonian is identical with the function $F(x^i, p_j)$ defined in (9.2.6). We wish to consider the null lines, for which L, and therefore also H, is zero. In view of the second set of (9.2.11), we see that

$$\dot{x}^k = a^{ik} p_i = \frac{\partial H}{\partial p_k}, \tag{9.2.13}$$

and this agrees with the first of (9.2.7), when σ is taken as parameter.

To show that the second set (9.2.7 b) hold, we require a formula for the derivatives of $H(x^i, p_j)$ which, in mechanics, is found with the help of the Legendre transformation, but which can be established directly as follows. We have

$$2\frac{\partial H(x^i, p_k)}{\partial x^j} + 2\frac{\partial L(x^i, \dot{x}^k)}{\partial x^j} = \frac{\partial a^{kh}}{\partial x^j} p_k p_h + \frac{\partial a_{km}}{\partial x^j} \dot{x}^k \dot{x}^m$$

$$= \frac{\partial a^{kh}}{\partial x^j} a_{km} \dot{x}^m p_h + \frac{\partial a_{km}}{\partial x^j} a^{kh} p_h \dot{x}^m$$

$$= \frac{\partial}{\partial x^j} (a^{kh} a_{km}) \dot{x}^m p_h = \frac{\partial}{\partial x^j} (\delta_m^h) \dot{x}^m p_h = 0.$$

Thus

$$\frac{\partial H(x^i, p_k)}{\partial x^j} = -\frac{\partial L(x^i, \dot{x}^k)}{\partial x^j},$$

so that from (9.2.10) and (9.2.11) we have

$$\dot{p}_i \equiv \frac{d}{d\sigma}\left(\frac{\partial L}{\partial \dot{x}^i}\right) = \frac{\partial L}{\partial x^i} = -\frac{\partial H}{\partial x^i}. \tag{9.2.14}$$

Since F and H are identical, this agrees with (9.2.7 b) when σ is taken as parameter. We have therefore shown that the geodesic null lines and the bicharacteristics are one and the same.

Geodesic null lines are important in the theory of relativity, where they represent the paths of light pulses. Thus a light pulse is propagated along a bicharacteristic of the wave equation.

The characteristic surfaces may be regarded as generated by $(N-2)$-parameter families of the bicharacteristics, as we have noted. We now see that the characteristic surfaces are generated by the geodesic null lines of the Riemannian space. Any surface generated by an $(N-2)$-parameter family of geodesic null lines is, therefore, a characteristic surface.

There is a geodesic line in each direction at a given point Q. If the direction is null—if $ds = 0$ initially—the geodesic line is a null line through its entire length. The null directions at Q are determined by the vanishing of the fundamental metric form, which, in a suitable system of local Cartesian coordinates, can be written

$$ds^2 = a_{ik}\, dx^i dx^k = (dt)^2 - (dx^1)^2 - \ldots - (dx^{N-1})^2, \qquad (9.2.15)$$

where we have set $x^N = t$. The null directions are therefore tangent to the cone

$$(t-t_0)^2 = (x^1-x_0^1)^2 + \ldots + (x^{N-1}-x_0^{N-1})^2 \qquad (9.2.16)$$

with vertex at $Q = (x_0^1, \ldots, x_0^{N-1}, t_0)$. There is an $(N-2)$-parameter family of directions at Q which are tangent to the cone.

The geodesic null lines through Q, when prolonged, generate a cone which, at Q, is tangent to the cone (9.2.16). This null cone of Q, which must be a characteristic surface, is known as the characteristic cone with vertex Q. This cone divides the space into three regions, which, in the terminology of the theory of relativity, may be called the past, present, and future. The past and future regions are interior to the retrograde half-cone $(t < 0)$ and the direct half-cone $(t > 0)$, respectively. The present is the region lying outside the cone.

There is a corresponding classification of directions at a point, and therefore of surface elements according to their normals. We say that a displacement dx^i is *timelike* if

$$ds^2 = a_{ik}\, dx^i dx^k > 0,$$

as would be the case for a displacement of t alone in (9.2.15). A positive timelike displacement carries us into the future, a negative one into the past. On the other hand, if a displacement dx^i causes ds^2 to be negative, it is *spacelike*. Under a spacelike displacement we are carried to another point of the present. The classification of surface elements is based on that of the normal displacement to the surface. If the normal is timelike, the surface is said to be spacelike; thus

$$(\nabla \varphi)^2 = a^{ik}\, \frac{\partial \varphi}{\partial x^i}\, \frac{\partial \varphi}{\partial x^k} = a^{ik} a_{kh} \dot{x}^h a_{im} \dot{x}^m = a_{hm} \dot{x}^h \dot{x}^m > 0$$

holds for a spacelike surface. Conversely, a timelike surface element is one whose normal is spacelike, and for which the condition $(\nabla\varphi)^2 < 0$ holds. Thus the null directions divide the timelike from the spacelike directions, and the characteristic surfaces are intermediate to the spacelike and the timelike surfaces.

Exercise 1. Show that a spacelike surface (of the kind $t = $ constant) contains no null directions. *Hint*: Show $ds^2 = dt^2 - d\sigma^2$, where $d\sigma^2$ is the metric in the surface.

Exercise 2. Show that the transition from (9.2.10) to the Hamiltonian form can be made if a potential energy $V(x^i)$ is present, so that $L = T - V$, provided that (9.2.12) is replaced by the definition $H = p_i \dot{x}^i - L$.

Exercise 3. Verify that (9.2.7 b) is satisfied along the bicharacteristics, as a consequence of (9.2.5) and (9.2.7 a).

Exercise 4. Show that (9.2.16) is the equation of the characteristic cone in a suitable system of Riemannian coordinates.

9.3. Discontinuities and singularities. The exceptional nature of characteristic surfaces, and of the null directions, is further revealed by a consideration of the discontinuities which a solution of a hyperbolic equation may exhibit. We recall that an elliptic equation has no real characteristic surfaces, and that its solutions are smooth and regular. With hyperbolic equations the situation is quite different.

Let a solution function $u(P)$ of $L[u] = 0$ be continuous, and possess continuous first order derivatives, but suppose that the second order derivatives of u possess discontinuities across a certain surface S. We then say that $u(P)$ has a second order discontinuity across S. It can now be shown that S *must be a characteristic surface*. To establish this result, let us denote by $\varphi(x^i) = 0$ the equation of S, and as in § 5.4, let us use round brackets ( ) to denote a discontinuity across S. Now u, ∇u, and all inner or tangential derivatives of u are continuous across S, so that the only second derivative not continuous is $u_{\varphi\varphi}$. Since $L[u] = 0$ holds on both sides of S we have, writing $L[u]$ in the $\lambda^1, ..., \lambda^{N-1}, \lambda^N = \varphi(x^i)$ coordinate system of § 9.1,

$$0 = (L[u]) = Q(\varphi)(u_{\varphi\varphi}).\qquad(9.3.1)$$

If $(u_{\varphi\varphi}) \neq 0$, that is, if there is a discontinuity, then $Q(\varphi) = (\nabla\varphi)^2$ must vanish. Thus $S: \varphi = 0$ must be characteristic, as stated.

On the characteristic surface S, the equation $L[u] = 0$ is valid, and, since it involves only inner quantities, states a condition which they must satisfy. Let us examine the form of this condition by writing out the differential equation in the $\lambda^1,..., \lambda^{N-1}, \lambda^N = \varphi$ coordinate system. Since $a^{NN} = Q(\varphi) = 0$ in this system, we have, dropping primes in (9.1.6),

$$L[u] = 2a^{\rho N}u_{\rho N}+\beta^N u_N+a^{\rho\sigma}u_{\rho\sigma}+\beta^\rho u_\rho+cu = 0. \qquad (9.3.2)$$

Here the Greek indices are to be summed from 1 to $N-1$. The operation of differentiation along the normal is, in these coordinates, given by

$$\frac{\partial}{\partial\tau} = a^{ik}\varphi_k\frac{\partial}{\partial x^k} = a^{\rho N}\frac{\partial}{\partial\lambda^\rho}, \qquad (9.3.3)$$

and so (9.3.2) becomes

$$2\frac{\partial u_N}{\partial\tau}+\beta^N u_N+J = 0, \qquad (9.3.4)$$

where

$$J = a^{\rho\sigma}u_{\rho\sigma}+\beta^\rho u_\rho+cu \qquad (9.3.5)$$

contains only u and derivatives of u tangential to the surface. Thus the condition imposed by $L[u] = 0$ on S takes the form of the ordinary differential equation (9.3.4) for the transverse first derivative $u_N = u_\varphi = \partial u/\partial\varphi$. This condition holds, in fact, whether u_N is continuous or not. If u_N does happen to be discontinuous, while J is continuous, then from (9.3.4) we see that the discontinuity (u_N) satisfies

$$2\frac{\partial(u_N)}{\partial\tau}+\beta^N(u_N) = 0. \qquad (9.3.6)$$

In other words, $$(u_N) = (u_N)_0 \exp\left[-\frac{1}{2}\int\beta^N d\tau\right] \qquad (9.3.7)$$

is different from zero at every point of the characteristic ray if it is not zero at an initial point.

If now $(u_N) = 0$, but $(u_{NN}) \neq 0$, we may differentiate $L[u] = 0$ with respect to φ, and then set $\varphi = 0$. This leads to an equation for u_{NN} of the form (9.3.4), viz.,

$$2\frac{\partial u_{NN}}{\partial\tau}+\beta^N u_{NN}+J_1 = 0, \qquad (9.3.8)$$

where J_1 contains continuous inner quantities only. Thus, taking discontinuities across S, we find that the discontinuity (u_{NN}) satisfies a relation like (9.3.6), and therefore, as in (9.3.7),

$$(u_{NN}) = (u_{NN})_0 \exp\left[-\tfrac{1}{2}\int\beta^N d\tau\right]. \qquad (9.3.9)$$

This relation shows that if a solution has a second order discontinuity at one point, the discontinuity is present along that bicharacteristic through the point which lies in the characteristic surface $\varphi = 0$.

The loci of singularities of solutions of hyperbolic equations are also restricted to lie on characteristic surfaces. Suppose that $L[u] = 0$ has a solution of the form
$$u = U\varphi^\alpha \quad (\alpha < 0), \qquad (9.3.10)$$
U and φ being regular functions of the variables. Using the
$$\lambda^1, ..., \lambda^{N-1}, \lambda^N = \varphi$$
coordinate system, we find, after some calculation,
$$L[U\varphi^\alpha] = \alpha(\alpha-1)Q(\varphi)U\varphi^{\alpha-2}+\alpha\varphi^{\alpha-1}[2a^{N\rho}U_\rho+\beta^N U]+\varphi^\alpha L[U], \quad (9.3.11)$$
and this expression is to vanish since $U\varphi^\alpha$ is a solution. We multiply (9.3.11) by $\varphi^{2-\alpha}$ and let $\varphi \to 0$. It follows that
$$Q(\varphi) = (\nabla\varphi)^2 = 0, \qquad (9.3.12)$$
and so our claim is established. *Singularities of solutions occur on characteristic surfaces.*

We see, therefore, that hyperbolic equations have no solutions with point singularities such as those which are so useful in connexion with elliptic equations. This fact means that in the integration theory for hyperbolic equations different methods must be used.

Returning to (9.3.11), let us suppose that the family of surfaces $\varphi = c$ are all characteristic, so that $Q(\varphi) = (\nabla\varphi)^2$ is zero identically. The first term of (9.3.11) drops out altogether, and now, if we multiply by $\varphi^{1-\alpha}$ and let φ tend to zero, we find
$$2\frac{dU}{d\tau}+\beta^N U = 0. \qquad (9.3.13)$$

Thus the coefficient U of the singularity may not vanish at any point of the bicharacteristic tangent to the surface $\varphi = c$. In particular, if $\varphi = 0$ is the equation of the characteristic cone $\Gamma = s^2 = 0$ of a point Q, then the solution must become infinite at every point of the cone—provided $U(Q)$ is not zero. In this case we have a 'conical singularity'. Solutions of this kind correspond 'formally' to the fundamental solutions of elliptic equations.

As was shown in § 4.4, hyperbolic equations with constant coefficients have solutions which are Bessel functions, with argument depending upon the geodesic distance s. The characteristic cone has, as invariant equation, just $s(P, Q) = 0$, so these solutions are singular for $s = 0$, that is, at every point of the characteristic cone. For instance, the equation
$$u_{tt}-u_{xx}-u_{yy}-u_{zz} = 0$$

has the solution

$$s^{-2} = \Gamma^{-1} = [(t-t_0)^2-(x-x_0)^2-(y-y_0)^2-(z-z_0)^2]^{-1}.$$

Exercise 1. Show that discontinuities of the third or higher orders occur only on characteristic surfaces, and satisfy an equation (9.3.9).

Exercise 2. Find solutions of

(a) $u_{tt}-u_{xx}+k^2u = 0,$

(b) $u_{tt}-u_{xx}-u_{yy}+k^2u = 0,$

which have conical singularities. *Hint*: Find solutions depending only on geodesic distance.

9.4. The propagation of waves. Consider a normal hyperbolic equation of the special form

$$L[u] = \frac{\partial^2 u}{\partial t^2} - a^{\alpha\beta}\frac{\partial^2 u}{\partial x^\alpha \partial x^\beta}+b^\alpha\frac{\partial u}{\partial x^\alpha}+b\frac{\partial u}{\partial t}+cu = 0, \qquad (9.4.1)$$

in which the Greek indices α, β are summed from 1 to $N-1$. The $(N-1)$-square matrix of coefficients $a^{\alpha\beta}$ is positive definite, since the signature of the metric for the full N variables is $(1, N-1)$.

Solutions of this equation may be plotted in the space of the spacelike variables x^α, and regarded as varying with the time t. Thus the solutions represent waves which travel through the space, and it is the manner of propagation of these waves which now claims our attention. We first examine the characteristic surfaces from the space variable point of view. An equation $\varphi(x^\alpha, t) = 0$ of a characteristic surface may be solved for t in the form

$$t = \psi(x^\alpha), \qquad (9.4.2)$$

and, in view of (9.1.9) and of the sign of the above coefficients $a^{\alpha\beta}$, the condition of being characteristic becomes

$$a^{\alpha\beta}\frac{\partial\psi}{\partial x^\alpha}\frac{\partial\psi}{\partial x^\beta} = 1. \qquad (9.4.3)$$

Thus the spatial cross-sections of characteristic surfaces are given by (9.4.2), where $\psi(x^\alpha)$ satisfies the first order partial differential equation (9.4.3). As t varies, a family of surfaces $\psi(x^\alpha) = t$ is generated.

We shall now demonstrate that these surfaces are the possible wavefronts. Consider a disturbance emanating from a source and which, at a given instant, is confined to a certain region of space. Let us suppose also that the edge or front of this disturbance is characterized by a discontinuity of the second order or higher. From § 9.3, we see that the locus in space-time of this discontinuity is indeed a characteristic surface, of the form (9.4.2). Accordingly, the family of surfaces $\psi(x^\alpha) = t$ must represent the successive positions in space occupied by the wave-front.

The magnitude of the discontinuity is governed by (9.3.9), which shows that the wave disturbance is propagated along the bicharacteristic curves in space-time. Let us examine the equations of these curves when they are written in the space variables. Since we may write $\varphi = t - \psi(x^\alpha)$, and since $a^{NN} = 1$, the equation (9.2.2) for $i = N$ becomes

$$dt = d\tau,$$

where in (9.2.2) we have written $d\tau$ for the independent differential along the curve. Thus we may replace $d\tau$ by dt in the remaining equations (9.2.2) which then become

$$\dot{x}^\alpha = \frac{dx^\alpha}{dt} = a^{\alpha\beta} \frac{\partial\psi}{\partial x^\beta}. \tag{9.4.4}$$

Here we have taken account of the negative signs in (9.4.1) and the relation $\varphi = t - \psi$.

The equations $x^\alpha = x^\alpha(t)$ of a bicharacteristic in space-time define a motion of a fictitious particle in the space, the velocity vector being $\dot{x}^\alpha$. From (9.4.4) we see that the velocity is in the direction of the normal to the surface $\psi(x^\alpha) = t$. The normal, as here understood, is the normal in the positive definite Riemann metric

$$d\sigma^2 = a_{\alpha\beta}\, dx^\alpha dx^\beta \quad (a^{\alpha\beta}a_{\beta\gamma} = \delta^\alpha_\gamma) \tag{9.4.5}$$

of the space variables. The disturbance, represented by the particle, is propagated along a ray which is an orthogonal trajectory of the family of wave-fronts. Moreover, it is propagated at unit speed, since we may readily conclude from (9.4.3) and (9.4.4) that

$$v^2 = \left(\frac{d\sigma}{dt}\right)^2 = a_{\alpha\beta} \frac{dx^\alpha}{dt} \frac{dx^\beta}{dt} = a^{\gamma\delta} \frac{\partial\psi}{\partial x^\gamma} \frac{\partial\psi}{\partial x^\delta} = 1. \tag{9.4.6}$$

Since also

$$\dot{\psi} = \frac{\partial\psi}{\partial x^\alpha} \dot{x}^\alpha = a^{\alpha\beta} \frac{\partial\psi}{\partial x^\alpha} \frac{\partial\psi}{\partial x^\beta} = 1, \tag{9.4.7}$$

we verify that the wave-front advances in step with the disturbance propagated along the rays.

These orthogonal trajectories, the rays, are the spatial projections of the bicharacteristic curves in space-time; and they enjoy properties relative to the space metric which the bicharacteristics have with respect to the metric of space-time. The equations of the characteristic curves of (9.4.3) are (9.4.4), (9.4.7), and the further equation

$$\frac{dp_\alpha}{dt} = \frac{d}{dt}\left(\frac{\partial\psi}{\partial x^\alpha}\right) = -\frac{\partial}{\partial x^\alpha}(a^{\beta\gamma})\frac{\partial\psi}{\partial x^\beta}\frac{\partial\psi}{\partial x^\gamma}. \tag{9.4.8}$$

However, as the reader will verify, this equation is automatically satisfied if (9.4.3) and (9.4.4) hold. That is, the orthogonal trajectories or rays are identical with the characteristic curves of (9.4.3).

The rays are also geodesics of the space metric (9.4.5). Since the formal proof of this follows closely the similar proof given in § 9.2 for the bi-characteristics, we shall leave to the reader the task of verifying this fact. It is to be noted that since the spatial metric is positive definite, there can be no null geodesic lines in space. This also implies that the rays are never tangential to the wave-fronts.

A disturbance which originates at a point Q at time t_0 will at time t have spread over a geodesic sphere of centre Q with radius $t-t_0$. The 'leading edge' of the wave-front will be sharp if the derivative u across the wave-front (in space-time, across the characteristic surface) is discontinuous. According to (9.3.6) and (9.3.7), if the wave-front is initially sharp, it will remain so, the degree of sharpness being attenuated by the quantity β^N. At time t, the disturbance may be confined to the surface of the geodesic sphere (assuming the initial disturbance to have been an impulse lasting for an infinitesimal length of time), in which case the propagation of the waves is said to be clean-cut. This is true of the propagation of light or of sound waves, for instance. On the other hand, there need be no sharp 'trailing edge' of the wave; there may be a residual disturbance which dies down slowly. This residual wave is present in the case of waves propagated over a water surface, there being two spatial coordinates present. In the following chapter the exact form of these solutions will be found, and the circumstances under which clean-cut propagation may occur will be determined.

A wave-front of any kind may be considered to be a locus of points each of which is emitting a disturbance in the fashion just described. After the lapse of an interval of time t, these secondary disturbances will each have spread to occupy a geodesic sphere of radius t about their respective sources. The new wave-front of the entire disturbance will now be the envelope of these geodesic spheres, and so it is geodesically parallel to the old front, at a distance $\sigma = t$. According to this construction, which was suggested by Huygens as the mode of propagation of light, the wave-fronts are propagated parallel to themselves, and at unit speed (Fig. 10).

This secondary wave picture of the propagation leads, as we have just seen, to the same construction of the wave-fronts, and therefore of their orthogonal trajectories, as the ray picture described above. A clean-cut disturbance may be thought of as consisting of particles moving along the orthogonal rays, or as a wave propagated by means of secondary waves arising at each point. Thus there is a kind of duality between the two

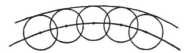

FIG. 10. Propagation of a wave-front.

pictures. In the classical theory of light there was no reason to prefer the wave picture on the basis of the propagation of the wave-fronts alone, and further experiments with diffraction were needed to decide the point. According to the quantum theory, neither picture by itself suffices to explain all of the known facts, and it must be supposed that the two points of view are each partial and complementary descriptions of a deeper and more subtle reality.

Exercise. Show that two wave-fronts $\psi_1 = t$, $\psi_2 = t$ which are tangent at one instant of time remain tangent for all time.

The wave propagation is particularly easy to visualize when the coefficients in the differential equation are constants. We shall here describe briefly the plane wave solutions of such equations. In § 4.4 it was seen that any normal hyperbolic equation with constant coefficients is equivalent to an equation of the special form

$$L[u] = \frac{\partial^2 u}{\partial t^2} - \sum_\alpha \frac{\partial^2 u}{(\partial x^\alpha)^2} + cu = 0, \qquad (9.4.9)$$

where c is, again, a constant. Let $\{a_\alpha\}$ ($\alpha = 1,..., N-1$) be components of a Cartesian space vector $\mathbf{a}$, and let

$$\mathbf{a} \cdot \mathbf{x} = a_\alpha x^\alpha$$

be the scalar product with the position vector $\mathbf{x}$. We look for solutions of the form
$$u = f(\mathbf{a} \cdot \mathbf{x} - t) = f(a_\alpha x^\alpha - t), \qquad (9.4.10)$$
which represent plane waves travelling in the direction of the vector $\mathbf{a}$. We find, as condition that (9.4.10) should be a solution,

$$\left\{ 1 - \sum_\alpha (a_\alpha)^2 \right\} f''(\mathbf{a} \cdot \mathbf{x} - t) + cf(\mathbf{a} \cdot \mathbf{x} - t) = 0. \qquad (9.4.11)$$

Two cases may now be distinguished. First, suppose that $c = 0$; then (9.4.11) holds if

$$\sum_\alpha (a_\alpha)^2 = 1$$

and if $f(\rho)$ is any function whatever having two derivatives. The velocity of the plane waves (9.4.10) is v, where

$$v^{-2} = \sum_\alpha (a_\alpha)^2,$$

so that the differential equation has plane wave solutions, of an arbitrary form, and travelling in an arbitrary direction, but always with unit velocity.

The case when $c \neq 0$ is quite different. We may choose any velocity different from unity, and then (9.4.11) becomes a differential equation for the function f. The two solutions of (9.4.11) are then sinusoidal or exponential, according as

$$\frac{1}{c}\left(1 - \sum_\alpha (a_\alpha)^2\right)$$

is positive or negative. In this case, therefore, the equation has plane wave solutions with arbitrary wave velocity (velocity 1 excepted) but of a specified form which depends on the wave velocity v. If we consider a solution which is a superposition of waves of this type, having various speeds, then the different components of the solution will travel at different rates, so that the form of the total wave will change with the time. This is the phenomenon of dispersion, which is therefore present whenever the coefficient of u in (9.4.9) is not zero.

Exercise. Show that solutions $u(x, t)$ of

$$u_{tt} - c^2 u_{xx} + (a+b)u_t + abu = 0$$

undergo dispersion unless $a = b$, in which case the most general solution has the form

$$u(x, t) = \exp[-at]\{f(x-ct) + g(x+ct)\}.$$

9.5. The initial value problem. In Chapter I, and again in Chapter VIII, we discussed initial value problems for equations of the normal hyperbolic type, and constructed solutions of these problems in certain special cases. Since the equation is of the second order in the time variable t, two initial conditions are required to determine a solution. These are usually given as the initial values of u and of its time derivative u_t. We shall now set out to prove that these initial conditions determine the solution uniquely in a certain region.

Let the differential equation take the form

$$L[u] = \frac{\partial^2 u}{\partial t^2} - a^{\alpha\beta}\frac{\partial^2 u}{\partial x^\alpha \partial x^\beta} + b^\alpha \frac{\partial u}{\partial x^\alpha} + b\frac{\partial u}{\partial t} + cu = 0, \qquad (9.5.1)$$

as in (9.4.1), so that the metric corresponding to the space variables x^α is positive definite. Let a solution u vanish, and have a vanishing first time derivative u_t, on a region S of the initial plane $t = 0$. We choose a point of general position P in space-time, with $t > 0$. We shall demonstrate that $u(P)$ is zero, provided that the retrograde characteristic cone C_P with vertex P cuts the initial plane inside the region S.

We start from the identity

$$2u_t[u_{tt} - a^{\alpha\beta}u_{x^\alpha x^\beta}] = \frac{\partial}{\partial t}(u_t)^2 + a^{\alpha\beta}\frac{\partial}{\partial t}(u_{x^\alpha}u_{x^\beta}) - 2a^{\alpha\beta}\frac{\partial}{\partial x^\alpha}(u_{x^\beta}u_t),$$

$$(9.5.2)$$

which is easily verified when the indicated differentiations are performed.

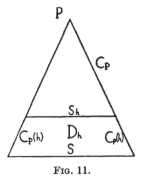

FIG. 11.

Let S_h denote that part of the plane $t = h$ which is cut off by the cone C_P (Fig. 11). We multiply $L[u]$ by the derivative u_t and integrate over the region D_h enclosed by S, S_h, and the cone C_P. Thus

$$2\int_{D_h} u_t L[u]\, dV = \int_{D_h}\left\{\frac{\partial}{\partial t}(u_t)^2 + \frac{\partial}{\partial t}(a^{\alpha\beta}u_{x^\alpha}u_{x^\beta}) - \frac{\partial}{\partial x^\alpha}(2a^{\alpha\beta}u_{x^\beta}u_t)\right\} dV +$$

$$+ \int_{D_h}\left\{b^\alpha u_{x^\alpha}u_t + bu_t^2 + cuu_t - \frac{\partial a^{\alpha\beta}}{\partial t}u_{x^\alpha}u_{x^\beta} + 2\frac{\partial a^{\alpha\beta}}{\partial x^\alpha}u_{x^\beta}u_t\right\} dV.$$

$$(9.5.3)$$

The first term on the right may be transformed to a surface integral by means of the divergence theorem. We find for it the expression

$$\int_{S + C_P(h) + S_h}\{u_t^2 n_t + a^{\alpha\beta}u_{x^\alpha}u_{x^\beta}n_t - 2a^{\alpha\beta}u_{x^\beta}u_t n_\alpha\}\, dS, \qquad (9.5.4)$$

where n_t and n_α are components of a vector normal to the surface. On the initial surface S the integrand vanishes by hypothesis. On the surface S_h, the normal vector may be taken to have $n_t = 1$, $n_\alpha = 0$ ($\alpha = 1,..., N-1$), and the contribution from S_h may be written

$$E(h) = \int_{S_h} \{u_t^2 + a^{\alpha\beta}u_{x^\alpha}u_{x^\beta}\} \, dS. \tag{9.5.5}$$

Note that $E(h)$ is a positive definite integral.

On the surface $C_P(h)$ of the cone, we have

$$n_t^2 = a^{\alpha\beta}n_\alpha n_\beta, \tag{9.5.6}$$

since the normal to this characteristic surface is of zero length. The contribution from $C_P(h)$ may therefore be written in the form

$$C(h) = \int_{C_P(h)} \frac{1}{n_t} \{a^{\alpha\beta}n_\alpha n_\beta u_t^2 - 2a^{\alpha\beta}u_{x^\beta}n_t u_t n_\alpha + a^{\alpha\beta}u_{x^\alpha}u_{x^\beta}n_t^2\} \, dS$$

$$= \int_{C_P(h)} \frac{a^{\alpha\beta}}{n_t} (u_t n_\alpha - u_{x^\alpha}n_t)(u_t n_\beta - u_{x^\beta}n_t) \, dS \geqslant 0, \tag{9.5.7}$$

since the quadratic form based on $a^{\alpha\beta}$ is positive definite, and since $n_t > 0$.

Let us denote by $-R$ the second integral on the right in (9.5.3). We see that R is a homogeneous quadratic integral over the unknown function u and its first partials u_{x^α} and u_t. From (9.5.3), (9.5.5), and (9.5.7) we find

$$0 = E(h) + C(h) - R,$$

whence, since $C(h)$ is non-negative,

$$E(h) \leqslant R. \tag{9.5.8}$$

This is an estimate of $E(h)$ in terms of R. The key to the success of this method is, that we can now find an estimate of R in terms of $E(h)$.

Since u vanishes on S ($t = 0$), we have

$$u(x^\alpha, t) = \int_0^t u_\tau(x^\alpha, \tau) \, d\tau,$$

and

$$u^2(x^\alpha, t) \leqslant t \int_0^t u_\tau^2(x^\alpha, \tau) \, d\tau,$$

by the Schwarz inequality. Thus, integrating over S_h,

$$\int_{S_h} u^2(x^\alpha, t) \, dS \leqslant h \int_{D_h} u_\tau^2(x^\alpha, \tau) \, dV \leqslant h \int_0^h E(k) \, dk,$$

in view of (9.5.5). Integrating once again from zero to h, we find

$$\int_{D_h} u^2 \, dV \leqslant h^2 \int_0^h E(k) \, dk.$$

Since　　　　$2|u_{x^\alpha} u_t| \leqslant u_{x^\alpha}^2 + u_t^2, \quad 2|uu_t| \leqslant u^2 + u_t^2,$

and since the coefficients of the quantities u, u_{x^α}, u_t in the integrand of R in (9.5.3) may be assumed to be bounded inside the cone C_P, we conclude that

$$R \leqslant K_1 \left[\int_0^h E(k) \, dk + \int_{D_h} u^2 \, dV \right]$$

$$\leqslant K_1(1+h^2) \int_0^h E(k) \, dk$$

$$\leqslant K \int_0^h E(k) \, dk \qquad \qquad \text{(for } h \leqslant h_0\text{)}.$$

Now we may employ (9.5.8) to gain our conclusion. We have

$$E(h) \leqslant K \int_0^h E(k) \, dk,$$

and, integrating from 0 to h, we find

$$\int_0^h E(h) \, dh \leqslant Kh \int_0^h E(k) \, dk.$$

But K is a fixed positive constant, and we can choose h so small that $Kh < 1$. We must then conclude that $E(k) = 0$ for $0 \leqslant k \leqslant h$. From (9.5.5) it follows that all derivatives of u vanish for $0 \leqslant t \leqslant h$, so that u vanishes there also. We may now repeat the whole process, starting with $t = h$, $t = 2h$,..., until the point P is reached. Thus $u(P) = 0$ as originally stated. *The solution of the initial value problem is unique.*

This proof can be modified to apply to any smooth spacelike initial surface, instead of the plane S ($t = 0$), the necessary changes being easily made.

Looking at this uniqueness theorem from a different point of view, we can say that it shows that the domain of dependence of the value of the solution at P is limited to the data of the initial value problem in the interior, and on the surface, of the retrograde null cone with vertex at P. Indeed, the value of $u(P)$ is independent of the data assigned outside of that portion S_P of the initial surface which is intercepted by the retrograde null cone. Thus no disturbance, or signal, can travel at faster than unit velocity in the space. Another statement of this fact is that the value of $u(P)$ can be influenced only by past events in space-time.

Exercise 1. Show that the initial value problem for $u_{xy} = 0$ may be incompatible if the initial line meets twice any line parallel to the x or y axes. State the corresponding facts for $u_{xx} - u_{yy} = 0$, $u_{xx} + u_{yy} - u_{tt} = 0$.

Exercise 2. Let Δ be the Laplacian associated with a positive definite metric on a spatial domain D, and consider the parabolic equation $u_t = \Delta u$, with boundary condition either $u(p, t) = 0$ or $u_n(p, t) = 0$. Show that $\int_D u^2 \, dV$ is a non-increasing function of t, and deduce that a solution of the initial value problem is unique.

Exercise 3. With the notation of the preceding exercise, consider the wave equation $u_{tt} = \Delta u$, with the boundary condition $u(p, t) = f(p)$, or $u_n(p, t) = 0$. Show that $\int_D [u_t^2 + (\nabla u)^2] \, dV$ is independent of t, and deduce a uniqueness theorem.

X

INTEGRATION OF THE WAVE EQUATION

THE integration of the linear elliptic equation, which was studied in detail in Chapter V, was based on Green's formula, and the fundamental solution with a point singularity. For hyperbolic equations Green's formula is still available; however, the fundamental solution no longer has a point singularity, but a conical singularity as described in § 9.3. Thus integrals of the fundamental solution over surfaces, or regions of space, are seen to be divergent, at least when $N > 2$. This is the cause of the difficulties encountered by Volterra and Hadamard, who nevertheless were able to carry out the integration with the aid of special techniques for evaluating divergent integrals.

The method to be described here was invented and perfected by Riesz, and is in fact applicable to elliptic equations as well. The outstanding feature of the Riesz method is the use of integrals of fractional order, a device which is most successful in circumventing the difficulties mentioned above. The order of these integrals will even be treated as a variable capable of assuming complex values. The solution for the initial value problem is found by a process of analytic continuation with respect to this parametric complex variable. Though this device may seem artificial, it will soon be evident that it is well adapted to the problem.

10.1. The Riemann–Liouville integral. We shall begin with the simplest type of integral of fractional order, the one-dimensional Riemann–Liouville integral

$$I^\alpha f(x) = \frac{1}{\Gamma(\alpha)} \int\limits_a^x (x-t)^{\alpha-1} f(t)\, dt. \qquad (10.1.1)$$

Our reason for studying this operator is twofold. We can develop a prototype of the full N-dimensional theory, 'in miniature', and also the formulae which we derive regarding (10.1.1) will find specific applications in the later work.

If we consider α to be a complex variable, the integral (10.1.1) converges for $R(\alpha) > 0$, and defines an analytic function of α. Thus $I^\alpha f(x)$ may be regarded as a function of the complex variable α and also as a functional depending on the values of $f(t)$ in the range from a to x. We now set down those properties of $I^\alpha f(x)$ which lead to its being regarded as an integral of fractional order α. First,

$$\frac{d}{dx} I^{\alpha+1} f(x) = \frac{d}{dx} \frac{1}{\Gamma(\alpha+1)} \int_a^x (x-t)^\alpha f(t) \, dt$$

$$= \frac{\alpha}{\Gamma(\alpha+1)} \int_a^x (x-t)^{\alpha-1} f(t) \, dt = I^\alpha f(x), \tag{10.1.2}$$

so that the operation of differentiation with respect to x lowers the index α by unity. Secondly, if n is a positive integer,

$$I^n f(x) = \frac{1}{(n-1)!} \int_a^x (x-t)^{n-1} f(t) \, dt = \int_a^x \ldots \int_a^t f(\tau) \, d\tau \ldots dt \tag{10.1.3}$$

is the n-fold indefinite integral of $f(x)$. This is clear for $n = 1$; for higher values of n it may be established directly, or by using (10.1.2) and noting that $I^\alpha f(a)$ is always zero.

The property of composition of the index α is

$$I^\alpha I^\beta f(x) = I^{\alpha+\beta} f(x). \tag{10.1.4}$$

To establish (10.1.4), we note that the left-hand side may be written

$$I^\alpha I^\beta f(x) = \frac{1}{\Gamma(\alpha)} \int_a^x (x-t)^{\alpha-1} \, dt \, \frac{1}{\Gamma(\beta)} \int_a^t (t-s)^{\beta-1} f(s) \, ds$$

$$= \frac{1}{\Gamma(\alpha)\Gamma(\beta)} \int_a^x f(s) \, ds \int_s^x (x-t)^{\alpha-1} (t-s)^{\beta-1} \, dt.$$

With $(x-s)\tau = x-t$, the inner integral is equal to

$$(x-s)^{\alpha+\beta-1} \int_0^1 \tau^{\alpha-1} (1-\tau)^{\beta-1} \, d\tau = \frac{(x-s)^{\alpha+\beta-1} \Gamma(\alpha)\Gamma(\beta)}{\Gamma(\alpha+\beta)}.$$

Thus $I^\alpha I^\beta f(x) = \dfrac{1}{\Gamma(\alpha+\beta)} \displaystyle\int_a^x (x-s)^{\alpha+\beta-1} f(s) \, ds = I^{\alpha+\beta} f(x)$

and (10.1.4) stands established.

Provided that $f(x)$ can be differentiated, the region of definition in the

α-plane of $I^\alpha f(x)$ can be extended to the left. Let us integrate by parts the integral

$$I^\alpha f(x) = \frac{1}{\Gamma(\alpha)} \int\limits_a^x f(t)(x-t)^{\alpha-1}\,dt = \frac{-1}{\alpha\Gamma(\alpha)} \int\limits_a^x f(t)\,d(x-t)^\alpha$$

$$= -\frac{(x-t)^\alpha f(t)}{\Gamma(\alpha+1)}\Big]_a^x + \frac{1}{\Gamma(\alpha+1)} \int\limits_a^x (x-t)^\alpha f'(t)\,dt$$

$$= \frac{(x-a)^\alpha f(a)}{\Gamma(\alpha+1)} + I^{\alpha+1}f'(x). \tag{10.1.5}$$

Repeating this process n times, we find

$$I^\alpha f(x) = \sum_{k=0}^{n-1} \frac{(x-a)^{\alpha+k}f^{(k)}(a)}{\Gamma(\alpha+k+1)} + I^{\alpha+n}f^{(n)}(x). \tag{10.1.6}$$

The right-hand side of (10.1.6) is analytic provided that $R(\alpha) > -n$, and so must constitute an analytic continuation of $I^\alpha f(x)$ into the left half-plane. The reader will recall that the gamma function $\Gamma(\alpha)$ is defined for $R(\alpha) < 0$ by exactly the same device: analytic continuation effected by integration by parts.

In formula (10.1.5), let $\alpha \to 0$. We find that

$$I^0 f(x) = f(a) + I^1 f'(x)$$

$$= f(a) + \int\limits_a^x f'(t)\,dt = f(x). \tag{10.1.7}$$

Thus the operator I^0 is the identity; it reproduces the function $f(x)$. From (10.1.6) we may, by a similar limiting process, calculate $I^{-n}f(x)$, where n is a positive integer. Since $\Gamma(\alpha+k+1)$ has a pole when $\alpha = -n$, $0 \leqslant k \leqslant n-1$, the finite sum in (10.1.6) vanishes as $\alpha \to -n$, and we have

$$I^{-n}f(x) = \lim_{\alpha \to -n} I^{\alpha+n}f^{(n)}(x) = I^0 f^{(n)}(x) = f^{(n)}(x). \tag{10.1.8}$$

The operator I^{-n} represents n successive differentiations.

When $\alpha = 0$ or $\alpha = -n$, the value of $I^\alpha f(x)$ no longer depends upon the range of values of the function $f(t)$ from a to x, but only on the values of $f(x)$ itself, and of the derivatives of $f(x)$ at the point x. For these values of α, I^α is a local operator, having values which depend only on the behaviour of $f(x)$ in the neighbourhood of the point x. This exceptional property of the negative integral values of I^α is the key to the mathematical aspect of the clean-cut wave propagation which was discussed in § 9.4, and which will be met with again in § 10.4.

Exercise 1. Show that $I^0f(x) = \lim_{\alpha \to 0} I^\alpha f(x) = f(x)$ if $f(x)$ is only continuous.

Exercise 2. Find explicitly the Riemann–Liouville integral of $f(t) = t^s$.

Exercise 3. How far to the left may $I^\alpha f(x)$ be defined if $f(x)$ is exactly n times differentiable?

Exercise 4. Verify the Euler beta-function formula

$$B(\alpha, \beta) = \int_0^1 \tau^{\alpha-1}(1-\tau)^{\beta-1}\,d\tau = \frac{\Gamma(\alpha)\Gamma(\beta)}{\Gamma(\alpha+\beta)}$$

by expressing $\Gamma(\alpha)\Gamma(\beta)$ as an iterated integral.

Exercise 5. Show that $B(\alpha, \alpha) = \tfrac{1}{2}B(\alpha, \tfrac{1}{2})$, and deduce the duplication formula $2^{1-2\alpha}\Gamma(\tfrac{1}{2})\Gamma(2\alpha) = \Gamma(\alpha)\Gamma(\alpha+\tfrac{1}{2})$.

Exercise 6. Assuming that $f(x) \sim e^{-|x|}$ as $x \to -\infty$, let $a \to -\infty$ in (10.1.1), and investigate the changes necessary in the formulae (10.1.3) to (10.1.7).

10.2. The fractional hyperbolic potential. Consider the wave-equation in N variables with constant coefficients (with $x^0 \equiv t$):

$$\Delta u = \frac{\partial^2 u}{\partial t^2} - \frac{\partial^2 u}{(\partial x^1)^2} - \cdots - \frac{\partial^2 u}{(\partial x^{N-1})^2} = \sum_{k=0}^{N-1} \epsilon_k \frac{\partial^2 u}{(\partial x^k)^2} = 0, \quad (10.2.1)$$

where $\epsilon_0 = 1$, $\epsilon_1 = \epsilon_2 = \ldots = \epsilon_{N-1} = -1$. The Lorentz metric corresponding to this equation is

$$ds^2 = dt^2 - (dx^1)^2 - \ldots - (dx^{N-1})^2 = \sum \epsilon_k (dx^k)^2, \quad (10.2.2)$$

and, since the underlying space is flat, the geodesic distance is given by

$$\Gamma(P, Q) = s^2(P, Q) = r^2(P, Q) = \sum_k \epsilon_k(x^k - \xi^k)^2, \quad (10.2.3)$$

where x^k, ξ^k are the coordinates of P and Q, respectively, in a suitable system of Lorentzian coordinates. The characteristic cone with vertex P has the equation $r(P, Q) = 0$, and the retrograde half-cone is characterized by the further condition $x^0 - \xi^0 > 0$. Let D^P denote the interior of the retrograde cone with vertex at P.

We now define the fractional integral which stands in the same relation to the operator Δ in (10.2.1) as does the Riemann–Liouville integral (10.1.1) to the operation d/dx. Let

$$I^\alpha f(P) = \frac{1}{H_N(\alpha)} \int_{D^P} r^{\alpha-N}(P, Q)f(Q)\,dV_Q, \quad (10.2.4)$$

where the numerical coefficient $H_N(\alpha)$ remains to be determined. We shall see that $H_N(\alpha)$ can be chosen so that both the property of differentiation

$$\Delta I^{\alpha+2}f(P) = I^{\alpha}f(P), \qquad (10.2.5)$$

and the composition law

$$I^{\alpha}I^{\beta}f(P) = I^{\alpha+\beta}f(P), \qquad (10.2.6)$$

are satisfied by the operation I^{α}. We shall assume that $f(P)$ tends to zero rapidly enough at infinity so that the integral converges if $R(\alpha)$ is large and positive. However, we must examine the convergence of the integral when $R(\alpha) < N$, since r vanishes on the cone, creating a singularity in the integrand. Suppose, for simplicity, that P is the origin of coordinates, and set

$$v^2 = \sum_{i=1}^{N-1} (\xi^i)^2 \qquad (v > 0)$$

so that $r^2(P, Q) = (\xi^0)^2 - v^2 = (\xi^0 + v)(\xi^0 - v).$

On the retrograde cone the factor $\xi^0 + v$ vanishes, and so the singularity of the integrand is given by

$$(\xi^0 + v)^{\frac{1}{2}(\alpha-N)}.$$

Taking v as one of the variables of integration, we see that the integral will converge with

$$\int_{-\xi^0} (\xi^0 + v)^{\frac{1}{2}(\alpha-N)} \, dv,$$

that is, if $R(\alpha) > N - 2$. For $R(\alpha) > N - 2$, $I^{\alpha}f(P)$ defined by (10.2.4) is a regular analytic function of the complex variable α.

We now consider the differentiation formula (10.2.5). The kernel of the integral operator I^{α} is

$$V^{\alpha}(P, Q) = \frac{r^{\alpha-N}(P, Q)}{H_N(\alpha)}, \qquad r^2 = \sum_k \epsilon_k(x^k - \xi^k)^2; \qquad (10.2.7)$$

and we shall choose $H_N(\alpha)$ so that

$$\Delta V^{\alpha+2}(P, Q) = V^{\alpha}(P, Q). \qquad (10.2.8)$$

Now $\dfrac{\partial}{\partial \xi^k} r^{\alpha+2-N} = (\alpha+2-N)r^{\alpha-N}\epsilon_k(x^k - \xi^k),$

and $\dfrac{\partial^2}{(\partial \xi^k)^2} r^{\alpha+2-N} = (\alpha+2-N)[r^{\alpha-N}\epsilon_k + (\alpha-N)r^{\alpha-N-2}\epsilon_k^2(x^k - \xi^k)].$

Therefore

$$\Delta r^{\alpha+2-N} = \sum_k \epsilon_k \frac{\partial^2 r^{\alpha+2-N}}{(\partial \xi^k)^2}$$

$$= (\alpha+2-N)[Nr^{\alpha-N} + (\alpha-N)r^{\alpha-N-2}\sum \epsilon_k(x^k - \xi^k)^2]$$

$$= (\alpha+2-N)\alpha r^{\alpha-N}.$$

We shall choose $H_N(\alpha)$ so that

$$\frac{(\alpha+2-N)\alpha}{H_N(\alpha+2)} = \frac{1}{H_N(\alpha)}, \tag{10.2.9}$$

then (10.2.8) is satisfied since

$$\Delta V^{\alpha+2}(P,Q) = \frac{\Delta r^{\alpha+2-N}}{H_N(\alpha+2)} = \frac{(\alpha+2-N)\alpha}{H_N(\alpha+2)} r^{\alpha-N}$$

$$= \frac{r^{\alpha-N}}{H_N(\alpha)} = V^{\alpha}(P,Q).$$

A solution of the difference equation (10.2.9) is

$$H_N(\alpha) = c(\alpha)2^{\alpha}\Gamma\left(\frac{\alpha}{2}\right)\Gamma\left(\frac{\alpha+2-N}{2}\right) \qquad (c(\alpha) = c(\alpha+2)).$$

The periodic function $c(\alpha)$ can be determined if we require that

$$I^{\alpha}e^{t} = e^{t} \qquad (t \equiv x^0), \tag{10.2.10}$$

a condition which is certainly consistent with (10.2.5). Since

$$I^{\alpha}e^{t} = \frac{1}{H_N(\alpha)} \int\limits_{D^P} e^{\xi^0}\left(\sum_k \epsilon_k(x^k-\xi^k)^2\right)^{\frac{1}{2}(\alpha-N)} dV,$$

we have, setting $\eta^k = \xi^k - x^k$, $\eta^0 = \xi^0 - t$,

$$I^{\alpha}e^{t} = \frac{e^{t}}{H_N(\alpha)} \int\limits_{D^P} e^{\eta^0}\left(\sum_k \epsilon_k(\eta^k)^2\right)^{\frac{1}{2}(\alpha-N)} dV_{\eta} = e^{t},$$

which implies that

$$H_N(\alpha) = \int\limits_{D^P} e^{\eta^0}\left(\sum_k \epsilon_k(\eta^k)^2\right)^{\frac{1}{2}(\alpha-N)} dV_{\eta}.$$

To evaluate this integral, set

$$\eta^0 = -r\cosh\theta, \qquad v^2 = \sum_{k=1}^{N-1}(\eta^k)^2 = r\sinh\theta.$$

We then find

$$H_N(\alpha) = \omega_{N-1}\int\limits_0^{\infty}\sinh^{N-2}\theta\, d\theta\int\limits_0^{\infty} e^{-r\cosh\theta}r^{\alpha-1}\, dr$$

$$= \pi^{\frac{1}{2}(N-2)}2^{\alpha-1}\Gamma\left(\frac{\alpha}{2}\right)\Gamma\left(\frac{\alpha+2-N}{2}\right). \tag{10.2.11}$$

Returning to the fractional integral (10.2.4), we see that if $R(\alpha) > N$, the kernel $V^{\alpha}(P,Q)$ is well-behaved on the cone, and

$$\Delta I^{\alpha+2}f(P) = \Delta\int\limits_{D^P} V^{\alpha+2}(P,Q)f(Q)\, dV_Q = \int\limits_{D^P} \Delta V^{\alpha+2}(P,Q)f(Q)\, dV_Q$$

$$= \int\limits_{D^P} V^{\alpha}(P,Q)f(Q)\, dV_Q = I^{\alpha}f(P).$$

This shows that (10.2.5) holds for $R(\alpha) > N$. But (10.2.5) is a relation between two analytic functions of the complex variable α, and so must be true whenever $I^{\alpha+2}f(P)$ and $I^{\alpha}f(P)$ are defined. That is, for all values of α, for which $I^{\alpha}f(P)$ is defined, we may differentiate formally under the integral sign.

We are now able to establish the composition formula (10.2.6). The left-hand side is, for $R(\alpha) > N$, $R(\beta) > N$,

$$I^{\alpha}I^{\beta}f(P) = \frac{1}{H_N(\alpha)} \int_{D^P} r^{\alpha-N}(P, Q) \, dV_Q \frac{1}{H_N(\beta)} \int_{D^Q} r^{\beta-N}(Q, R)f(R) \, dV_R$$

$$\tag{10.2.12}$$

$$= \frac{1}{H_N(\alpha)H_N(\beta)} \int_{D^P} f(R) \, dV_R \int_{D_R^P} r^{\alpha-N}(P, Q)r^{\beta-N}(Q, R) \, dV_Q.$$

In changing the order of integration, we observe that R ranges over all points in D^Q, where Q lies in D^P, so R ranges over D^P. With R fixed, Q

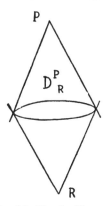

FIG. 12. The double cone.

ranges over those points of D^P such that R lies in the retrograde cone of Q; that is, so that Q lies in the forward or direct cone of R. The double cone D_R^P therefore denotes the volume enclosed by the retrograde cone of P and the direct cone of R (Fig. 12).

The inner integral on the right in (10.2.12) contains only invariant quantities, and is of spatial dimension $\alpha+\beta-N$ since the volume element has dimension N. If O and O' denote points whose invariant distance is $r(O, O') = 1$, then

$$\int_{D_R^P} r^{\alpha-N}(P, Q)r^{\beta-N}(Q, R) \, dV_Q$$

$$= r^{\alpha+\beta-N}(P, R) \int_{D_{O'}^O} r^{\alpha-N}(O, Q')r^{\beta-N}(Q', O') \, dV_{Q'}.$$

This last integral depends only on α and β, and may be denoted by $B_N(\alpha, \beta)$. From (10.2.12) we have, therefore,

$$I^\alpha I^\beta f(P) = \frac{B_N(\alpha, \beta)}{H_N(\alpha)H_N(\beta)} \int_{D^P} r^{\alpha+\beta-N}(P, R)f(R)\, dV_R$$

$$= \frac{B_N(\alpha, \beta)H_N(\alpha+\beta)}{H_N(\alpha)H_N(\beta)} I^{\alpha+\beta}f(P).$$

However, since we know that $I^\alpha I^\beta e^t = e^t = I^{\alpha+\beta}e^t$, the coefficient of $I^{\alpha+\beta}f(P)$ in this formula must be equal to unity; that is,

$$B_N(\alpha, \beta)H_N(\alpha+\beta) = H_N(\alpha)H_N(\beta),$$

and so $\qquad\qquad I^\alpha I^\beta f(P) = I^{\alpha+\beta}f(P),$

for all functions $f(P)$. Therefore (10.2.6) is proved.

Exercise. Show that

$$I^\alpha \exp[a_0 t + a_1 x^1 + \ldots + a_{N-1} x^{N-1}]$$
$$= (a_0^2 - a_1^2 - \ldots - a_{N-1}^2)^{-\frac{1}{2}\alpha} \exp[a_0 t + \ldots + a_{N-1} x^{N-1}].$$

10.3. The Cauchy problem. The initial value problem we wish to solve is that of calculating the solution $u(P)$ of

$$\Delta u(P) = f(P), \qquad\qquad (10.3.1)$$

where $\qquad\qquad \dfrac{\partial u(p)}{\partial n} = g(p), \qquad u(p) = h(p) \qquad\qquad (10.3.2)$

are assigned on a spacelike initial surface S (*frontispiece*). We have seen, in § 9.5, that the solution is unique, and that $u(P)$ depends only on the values of $f(P)$ inside the retrograde cone D_S^P above the surface S; and on the values of $g(p)$ and $h(p)$ on the portion S^P of S cut by the cone C^P with vertex P itself. To set up a formula for the solution of this problem, we shall apply Green's theorem to the region D_S^P, taking as one of the argument functions the kernel $V^\alpha(P, Q)$ of the fractional integral $I^\alpha f(P)$. If $R(\alpha) > N+1$, this function, and its first derivatives, vanish on the surface C^P of the cone, and we may therefore omit the surface integrals over the cone C^P. Under these circumstances Green's formula may be written

$$\int_{D_S^P} (u\, \Delta v - v\, \Delta u)\, dV = -\int_{S^P} \left(u \frac{\partial v}{\partial n} - v \frac{\partial u}{\partial n}\right) dS, \qquad (10.3.3)$$

the minus sign being inserted on the right because it is convenient to take the normal on S^P in the forward direction—facing into D_S^P.

Setting $\qquad v = v(Q) = V^{\alpha+2}(P, Q) = \dfrac{r^{\alpha+2-N}(P, Q)}{H_N(\alpha)},$

and noting (10.2.8), we see that Green's formula reads

$$\int\limits_{D_S^P} \left(u(Q)V^{\alpha}(P, Q) - V^{\alpha+2}(P, Q)\, \Delta u(Q) \right) dV_Q$$

$$= + \int\limits_{S^P} \left(\frac{\partial u(q)}{\partial n_q} V^{\alpha+2}(P, q) - u(q) \frac{\partial V^{\alpha+2}(P, q)}{\partial n_q} \right) dS_q. \qquad (10.3.4)$$

From (10.2.4) we see that the first term on the left is $I^{\alpha}u(P)$, and the second term $I^{\alpha+2}\Delta u(P)$. Thus (10.3.4) may be written

$$I^{\alpha}u(P) = I^{\alpha+2}\Delta u(P) + V^{\alpha+2}(P) + W^{\alpha+2}(P), \qquad (10.3.5)$$

where the surface layer potentials V^{α} and W^{α} are defined by

$$V^{\alpha}(P) = \int\limits_{S^P} V^{\alpha}(P, q) \frac{\partial u(q)}{\partial n_q}\, dS_q \qquad (10.3.6)$$

and $\qquad\qquad W^{\alpha}(P) = - \int\limits_{S^P} \frac{\partial V^{\alpha}(P, q)}{\partial n_q} u(q)\, dS_q. \qquad (10.3.7)$

These correspond to the single and double layers, respectively, in elliptic potential theory.

Formula (10.3.5) shall be the basis of our method of solution. We will show that the various potentials appearing in the formula can be continued analytically to the left of $R(\alpha) = N - 2$, and that the analytical continuation of $I^{\alpha}u(P)$, which we shall denote by the same symbol, satisfies

$$I^0 u(P) = u(P). \qquad (10.3.8)$$

If, then, the volume potential of density $f(P)$, and the single and double layers of density $g(p)$ and $h(p)$ are constructed, and continued analytically to $\alpha \to 0$, we should arrive at a formal solution of the initial value problem

$$u(P) = I^2 f(P) + V^2(P) + W^2(P). \qquad (10.3.9)$$

Our task is now to show that this continuation process can be performed, and that it leads to an actual solution.

Exercise. Show that

$$\Delta V^{\alpha+2}(P) = V^{\alpha}(P), \qquad I^{\alpha}V^{\beta}(P) = V^{\alpha+\beta}(P),$$

$$\Delta W^{\alpha+2}(P) = W^{\alpha}(P), \quad \text{and} \quad I^{\alpha}W^{\beta}(P) = W^{\alpha+\beta}(P).$$

10.4. Verification of the solution. In order to show that the fractional hyperbolic potentials have the properties requisite for the solution of the Cauchy problem, we must study them in some detail. The first need is a coordinate system adapted to the problem. We shall, in the following, assume that $N > 2$, leaving the slight modifications necessary when $N = 2$ to the reader.

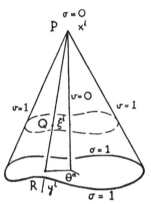

Fɪɢ. 13. A coordinate system.

Let P be the vertex of the cone C^P, D_S^P the region enclosed by C and the spacelike surface S^P. Let P have Lorentzian coordinates x^i, and let R be a point of S^P with coordinates y^i (Fig. 13). Let

$$v^2 = \frac{\sum_{i=1}^{N-1}(x^i-y^i)^2}{(x^0-y^0)^2} \quad (v > 0), \tag{10.4.1}$$

so that $v = 1$ on the cone C where $r(P, R) = 0$, and $v = 0$ on the line through P parallel to the x^0-axis. Let $r_0 = x^0-y^0$ denote the 'time interval' between P and R. Then

$$r(P, R) = r_0(1-v^2)^{\frac{1}{2}}. \tag{10.4.2}$$

Let Q be a point on the ray PR with coordinates ξ^i and let the ratio of $r(P, Q)$ to $r(P, R)$ be denoted by σ: $\xi^i = x^i+\sigma r^i$, $r^i = (y^i-x^i)$;

$$r(P, Q) = \sigma r(P, R) = \sigma r_0(1-v^2)^{\frac{1}{2}}. \tag{10.4.3}$$

Values of σ for points Q on the cone C can be defined by continuity. Then the surface S has the equation $\sigma = 1$. Together with v and σ, we shall choose $N-2$ angular coordinates $\theta^1,..., \theta^{N-2}$ to specify directions in the surface S. Then v, σ, and the θ^α ($\alpha = 1,..., N-2$), form a coordinate system in D_S^P.

We require the expressions for volume and surface elements in these coordinates. In the hyperplane

$$\tau = x^0 - \xi^0 = \text{constant}$$

the element of 'area' may be written down in spherical polar coordinates $v, \theta^1, ..., \theta^{N-2}$, with $\rho = v\tau$ as radial variable:

$$dA = \tau^{N-1} v^{N-2}\, dv d\Omega.$$

Here $d\Omega$ is the element of solid angle of the angular variables θ^α. The full volume element is

$$dV = d\xi^0 dA = r_0\, d\sigma (r_0 \sigma)^{N-1} v^{N-2}\, dv d\Omega$$
$$= r_0^N\, \sigma^{N-1} v^{N-2}\, d\sigma dv d\Omega. \tag{10.4.4}$$

The surface element dS on S^P may be found from dA by allowing for the inclination of the normal $\mathbf{n}$ to S^P, of the x^0-axis. Since $\sigma = 1$ on S^P, we have

$$|\mathbf{r}(P, R) \cdot \mathbf{n}|\, dS = r\, dA$$

and so

$$dS = r_0^N |\mathbf{r}(P, R) \cdot \mathbf{n}|^{-1} v^{N-2}\, dv d\Omega. \tag{10.4.5}$$

In forming the double layer we shall need to calculate

$$\frac{\partial}{\partial n} r^2(P, R) = \nabla(r^2) \cdot \mathbf{n} = 2\mathbf{r} \cdot \mathbf{n} = -2|\mathbf{r} \cdot \mathbf{n}|,$$

the scalar product being negative since n_0 is positive and r_0 is negative. Thus

$$\frac{\partial}{\partial n}(r^2(P, R))\, dS = -2r_0^N\, v^{N-2}\, dv d\Omega. \tag{10.4.6}$$

Now consider the volume potential

$$I^\alpha f(P) = \frac{1}{H_N(\alpha)} \int\limits_{D_S^P} r^{\alpha-N}(P, Q) f(Q)\, dV_Q = \frac{2^{1-\alpha}}{\pi^{\frac{1}{2}(N-2)}} \frac{1}{\Gamma(\frac{1}{2}\alpha)} \frac{1}{\Gamma(\frac{1}{2}\alpha - \frac{1}{2}N + 1)} \times$$

$$\times \iiint f(Q)|r_0|^{\alpha-N}(1-v^2)^{\frac{1}{2}(\alpha-N)}\sigma^{\alpha-N}|r_0|^N \sigma^{N-1} v^{N-2}\, dv d\sigma d\Omega. \tag{10.4.7}$$

This integral can be written in the form

$$\frac{2^{1-\alpha}\sqrt{\pi}}{\Gamma(\frac{1}{2}N - \frac{1}{2})\Gamma(\frac{1}{2}\alpha - \frac{1}{2}N + 1)} \int\limits_0^1 F_1(\alpha, v^2, P) v^{N-2}(1-v^2)^{\frac{1}{2}(\alpha-N)}\, dv, \tag{10.4.8}$$

where the auxiliary function $F_1(\alpha, v^2, P)$ is defined by

$$F_1(\alpha, v^2, P) = \frac{2}{\omega_{N-1}\, \Gamma(\frac{1}{2}\alpha)} \int\limits_0^1 \sigma^{\alpha-1}\, d\sigma \int\limits_{\Omega_{N-1}} f(Q)|r_0|^\alpha\, d\Omega. \tag{10.4.9}$$

In the expression (10.4.8), set $\beta = \frac{1}{2}(\alpha - N + 2)$, and $t = v^2$; then

$$I^\alpha f(P) = \frac{2^{-\alpha}}{\Gamma(\frac{1}{2}N - \frac{1}{2})} \frac{\sqrt{\pi}}{\Gamma(\beta)} \int_0^1 F_1(\alpha, t, P) t^{\frac{1}{2}(N-3)}(1-t)^{\beta-1} \, dt. \quad (10.4.10)$$

Let us now regard α and β as separate variables, independent of each other. (It can be shown that this will not lead to any error in the final result.) Then the integral (10.4.10) has the form of a one-dimensional Riemann–Liouville integral of fractional order β, with $x = 1$ and $a = 0$, of the function $F_1(\alpha, t, P) t^{\frac{1}{2}(N-3)}$. According to (10.1.6), we can continue this function analytically into the left half-plane of the variable β, provided only that the integrand can be differentiated a certain number of times.

First let us study the continuation to $\alpha = 0$, and verify that $I^0 f(P) = f(P)$. The value of $F_1(\alpha, v^2, P)$ must be calculated as a limit, since the expression (10.4.9) becomes indeterminate for $\alpha = 0$. Denoting the inner integral by $\omega_{N-1} \bar{f}(\sigma)$, we see that the limit must be taken as $\alpha \to 0$ of

$$\frac{2}{\Gamma(\frac{1}{2}\alpha)} \int_0^1 \sigma^{\alpha-1} \bar{f}(\sigma) \, d\sigma = \frac{2}{\Gamma(\frac{1}{2}\alpha)} \int_0^1 \sigma^{\alpha-1} [\bar{f}(0) + \sigma \bar{f}'(\xi)] \, d\sigma$$

$$= \frac{2}{2/\alpha + \ldots} \left[\bar{f}(0) \int_0^1 \sigma^{\alpha-1} \, d\sigma + \int_0^1 \sigma^\alpha \bar{f}'(\xi) \, d\sigma \right] = \bar{f}(0) + O(\alpha).$$

Thus the limit must be $f(0)$. But for $\sigma = 0$, the point Q coincides with P independently of the angular variables. The factor r_0^α may be set equal to unity in the limit, and we therefore conclude that

$$F_1(0, t, P) = f(P).$$

The integrand in (10.4.10) now takes the form

$$f(P) t^{\frac{1}{2}(N-3)}(1-t)^{\beta-1},$$

and it follows that the expression for $I^\alpha f(P)$ becomes

$$\frac{2^{-\alpha}}{\Gamma(\frac{1}{2}N - \frac{1}{2})} \frac{\sqrt{\pi}}{\Gamma(\beta)} f(P) \mathrm{B}\left(\frac{N-1}{2}, \beta\right) = \frac{2^{-\alpha}\sqrt{\pi}}{\Gamma(\beta + \frac{1}{2}N - \frac{1}{2})} f(P) \to f(P),$$

when we set $\alpha = 0$, $\beta = -\frac{1}{2}(N+2)$. Thus finally,

$$I^0 f(P) = f(P). \quad (10.4.11)$$

The Riesz volume potential of zero order is the identity operator.

Next we must study the properties of $I^2 f(P)$. Since $\beta = \frac{1}{2}(\alpha - N + 2)$, we see that $\beta \to -\frac{1}{2}(N-4)$ as $\alpha \to 2$. Thus if N is even, $N \geqslant 4$, β tends to a (negative) integer, while if N is odd, the value of β is half an odd integer. Now, according to (10.4.10), $I^\alpha f(P)$ is given by a one-dimensional Riemann–Liouville integral of order β. From (10.1.8) we recall that if β is equal to zero or a negative integer, the Riemann–Liouville integral is a local operator, depending only on the values of its integrand at the argument point x, which in (10.4.10) is equal to unity. Therefore, if N is even, $I^2 f(P)$ can depend only on values of $F_1(\alpha, t, P)$ and so of $f(Q)$, for which $t = 1$; that is, only values of $f(Q)$ on the null cone C^P. *In a space of even total dimension, the value of $I^2 f(P)$ depends only on values of $f(Q)$ on the cone C^P.*

It is at this point and in this way that the mathematical formalism reveals that the propagation of waves in an even-dimensional Lorentzian space is clean-cut. For N odd, the 'local' property of the Riemann–Liouville integral does not enter for $\alpha = 2$; and the values of $f(P)$ from the interior of the cone contribute to a non-zero residual wave.

From (10.2.5) we see that

$$\Delta I^2 f(P) = I^0 f(P) = f(P),\qquad(10.4.12)$$

so that the potential $I^2 f(P)$ is a solution of the non-homogeneous differential equation (10.3.1). In connexion with the initial conditions (10.3.2), we note that, as P tends to the surface S, the difference $r_0 = x^0 - y^0$ tends to zero. Therefore, since

$$F_1(2, t, P) = \frac{2}{\omega_{N-1}} \int_0^1 \sigma \, d\sigma \int_{\Omega_{N-1}} f(Q)|r_0|^2 \, d\Omega = O(r_0^2),$$

we see that F_1, together with its first partial derivatives, tends to zero as $r_0 \to 0$. Hence the same is true of $I^2 f(P)$:

$$I^2 f(p) = 0, \qquad \frac{\partial I^2 f(p)}{\partial n} = 0.\qquad(10.4.13)$$

The volume potential has vanishing initial values.

Corresponding properties of the surface layer potentials may be found by a quite similar analysis. For the single layer $V^\alpha(P)$ we have

$$V^\alpha(P) = \frac{1}{H_N(\alpha)} \int_{S^P} g(q)\, r^{\alpha-N}(P, q)\, dS_q$$

$$= \frac{2^{1-\alpha}}{\pi^{\frac{1}{2}(N-2)}} \frac{1}{\Gamma(\frac{1}{2}\alpha)} \cdot \frac{1}{\Gamma(\frac{1}{2}\alpha - \frac{1}{2}N + 1)} \iint g(q)\, \frac{r_0^\alpha (1-v^2)^{\frac{1}{2}(\alpha-N)} v^{N-2}}{(\mathbf{r}(P,q)\cdot\mathbf{n})}\, dv d\Omega$$

$$= \frac{2^{2-\alpha}}{\Gamma(\frac{1}{2}N - \frac{1}{2})} \frac{\sqrt{\pi}}{\Gamma(\frac{1}{2}\alpha - \frac{1}{2}N + 1)} \int_0^1 F_2(\alpha, v^2, P) v^{N-2}(1-v^2)^{\frac{1}{2}(\alpha-N)}\, dv$$

$$(10.4.14)$$

where the auxiliary function $F_2(\alpha, v^2, P)$ is defined as

$$F_2(\alpha, v^2, P) = \frac{1}{\Gamma(\frac{1}{2}\alpha)\omega_{N-1}} \int_{\Omega_{N-1}} g(q)r_0^\alpha(\mathbf{r}\cdot\mathbf{n})^{-1}\,d\Omega. \qquad (10.4.15)$$

Again, we define $\beta = \frac{1}{2}(\alpha-N+2)$, and write $t = v^2$, so that $V^\alpha(P)$ takes the form

$$V^\alpha(P) = \frac{2^{1-\alpha}}{\Gamma(\frac{1}{2}N-\frac{1}{2})}\frac{\sqrt{\pi}}{\Gamma(\beta)}\int_0^1 F_2(\alpha, t, P)t^{\frac{1}{2}(N-3)}(1-t)^{\beta-1}\,dt. \qquad (10.4.16)$$

Since this is a Riemann–Liouville integral, the analytic continuation to values of β corresponding to $\alpha = 0$ or $\alpha = 2$ is possible, by (10.1.6). For $\alpha = 0$, we note that the reciprocal factor $\Gamma(\frac{1}{2}\alpha)$ in (10.4.15) leads to $F_2(0, v^2, P)$ being zero, since the integral in (10.4.15) remains finite for $\alpha = 0$. Therefore

$$V^0(P) = 0,$$

and it follows from $\Delta V^{\alpha+2}(P) = V^\alpha(P)$ that

$$\Delta V^2(P) = 0. \qquad (10.4.17)$$

When N is even, $\geqslant 4$, $\alpha = 2$ corresponds to $\beta = \frac{1}{2}(4-N)$ which is a negative integer or zero, and so the local property of the Riemann–Liouville integral comes into play in (10.4.16). The values of F_2 (and therefore of $g(q)$) for $t = 1$ only, enter into the value of $V^2(P)$. That is, $V^2(P)$ depends only on values of $g(q)$ on the rim s_P where the surface S^P and the null cone C^P intersect. *Thus the clean-cut wave property holds for the single layer also, provided that N is even and $\geqslant 4$.*

To examine the initial values of the single layer, let P tend to the initial surface S along the normal to S at the point p. Since S is spacelike, the normal direction is timelike, and a system of coordinates can be found in which the normal is a parametric line of the timelike coordinate variable x^0. For simplicity, let us assume that S is flat in the neighbourhood of p, and has the equation $x^0 = 0$. For the normal n to S, we now have $n^0 = 1$, $n^i = 0$ for $i \neq 0$, and, as $P \to p$,

$$F_2(2, t, P) = \frac{1}{\omega_{N-1}} \int_{\Omega_{N-1}} g(q)r_0^2\frac{1}{r_0}\,d\Omega \to 0, \qquad (10.4.18)$$

as $r_0 = x^0 \to 0$. Thus we conclude that

$$V^2(p) = 0. \qquad (10.4.19)$$

To determine the normal derivative on S, we calculate from (10.4.18),

$$\frac{\partial F_2(2, t, P)}{\partial n} = \frac{\partial F_2(2, t, P)}{\partial r_0} = \frac{1}{\omega_{N-1}} \int_{\Omega_{N-1}} g(q)\,d\Omega \to g(p)$$

as $r_0 \to 0$, since the values of $g(q)$ concerned are those lying in the portion S_P, which shrinks to the point p. Substituting this limit in (10.4.16) we see that

$$
\frac{\partial V^2(p)}{\partial n} = \frac{\sqrt{\pi}}{2\Gamma(\frac{1}{2}N-\frac{1}{2})} \left[\frac{1}{\Gamma(\beta)} \int_0^1 g(p) t^{\frac{1}{2}(N-3)}(1-t)^{\beta-1} \, dt \right]_{\beta=\frac{1}{2}(4-N)}
$$

$$
= \frac{\sqrt{\pi}}{2\Gamma(\frac{1}{2}N-\frac{1}{2})} g(p) \left[\frac{B(\frac{1}{2}N-\frac{1}{2},\beta)}{\Gamma(\beta)} \right]_{\beta=\frac{1}{2}(4-N)} = g(p).
$$

(10.4.20)

The single layer vanishes on S, and its normal derivative is equal to the density $g(p)$.

Turning now to the double layer $W^\alpha(P)$, we see from (10.3.7) and (10.4.6) that

$$
W^\alpha(P) = -\frac{\alpha-N}{2H_N(\alpha)} \int_{S^P} h(q) r^{\alpha-N-2} \frac{\partial r^2}{\partial n} \, dS_q
$$

$$
= \frac{2^{1-\alpha}}{\pi^{\frac{1}{2}(N-2)}} \frac{1}{\Gamma(\frac{1}{2}\alpha)} \frac{\alpha-N}{\Gamma(\frac{1}{2}\alpha-\frac{1}{2}N+1)} \int\int h(q) r_0^{\alpha-2} v^{N-2}(1-v^2)^{\frac{1}{2}(\alpha-N-2)} \, dv \, d\Omega
$$

$$
= \frac{2^{3-\alpha}\sqrt{\pi}}{\Gamma(\frac{1}{2}N-\frac{1}{2})} \frac{1}{\Gamma(\frac{1}{2}\alpha-\frac{1}{2}N)} \int_0^1 F_3(\alpha, v^2, P) v^{N-2}(1-v^2)^{\frac{1}{2}(\alpha-N-2)} \, dv,
$$

(10.4.21)

where the third auxiliary function $F_3(\alpha, v^2, P)$ is defined as

$$
F_3(\alpha, v^2, P) = \frac{1}{\Gamma(\frac{1}{2}\alpha)\omega_{N-1}} \int_{\Omega_{N-1}} h(q) r_0^{\alpha-2} \, d\Omega.
$$

(10.4.22)

The method of analytic continuation is again the same, with the slight difference that we now set

$$
\beta = \frac{\alpha-N}{2}.
$$

(10.4.23)

Thus, with $t = v^2$, we have

$$
W^\alpha(P) = \frac{2^{2-\alpha}\sqrt{\pi}}{\Gamma(\frac{1}{2}N-\frac{1}{2})} \frac{1}{\Gamma(\beta)} \int_0^1 F_3(\alpha, t, P) t^{\frac{1}{2}(N-3)}(1-t)^{\beta-1} \, dt.
$$

(10.4.24)

The values $\alpha = 0$ and $\alpha = 2$ correspond to $\beta = -\frac{1}{2}N$ and $\beta = \frac{1}{2}(2-N)$. Again, for N even, these values are negative integers and it follows that $W^2(P)$ depends only on values of F_3, or of $h(q)$, with $t = 1$. That is, only values of $h(q)$ on the rim s_P of the surface S^P affect the value of $W^2(P)$, when N is even. We note that *the clean-cut wave propagation property holds* *for the double layer when N is even; and is also valid for $N = 2$, since then $\alpha = 2$ corresponds to $\beta = 0$.*

For $\alpha = 0$, we see from (10.4.22) that $F_3(\alpha, t, q)$ is zero, due to the singularity of $\Gamma(\tfrac{1}{2}\alpha)$ in the denominator. Thus we have $W^0(P) = 0$ and, since $\Delta W^{\alpha+2}(P) = W^\alpha(P)$ holds for the analytic function $W^\alpha(P)$,

$$\Delta W^2(P) = W^0(P) = 0. \tag{10.4.25}$$

Regarding the initial values, we see from (10.4.22) that

$$F_3(2, t, P) = \frac{1}{\omega_{N-1}} \int\limits_{\Omega_{N-1}} h(q)\, d\Omega \to h(p) \tag{10.4.26}$$

as $P \to p$ on S. From (10.4.24) we then have

$$W^2(P) = \frac{\sqrt{\pi}}{\Gamma(\tfrac{1}{2}N - \tfrac{1}{2})} \left[\frac{1}{\Gamma(\beta)} \int\limits_0^1 h(p) t^{\tfrac{1}{2}(N-3)}(1-t)^{\beta-1}\, dt \right]_{\beta=\frac{1}{2}(2-N)}$$

$$= \frac{\sqrt{\pi}}{\Gamma(\tfrac{1}{2}N - \tfrac{1}{2})} h(p) \left[\frac{B(\tfrac{1}{2}N - \tfrac{1}{2}, \beta)}{\Gamma(\beta)} \right]_{\beta=\frac{1}{2}(2-N)} = h(p). \tag{10.4.27}$$

To calculate the normal derivative of $W^2(p)$, let us again suppose that the equation of S is $x^0 = 0$; that is, we shall neglect any curvature of the surface. As P approaches its limit p along a normal to S, we note that in the integrand of (10.4.26) there enter values of $h(q)$ on a sphere (circle if $N = 3$) of radius $r_0 v$ with centre at p. Let y^i ($i = 1, ..., N-1$) be Cartesian coordinates on S, and let λ^i be the direction cosine of the y^i-axis with the ray from p to the variable point q. Then

$$\frac{\partial F_3(2, t, P)}{\partial n} = \frac{\partial F_3(2, t, P)}{\partial r_0} = \frac{v}{\omega_{N-1}} \int\limits_{\Omega_{N-1}} \frac{\partial h(q)}{\partial y^i} \lambda^i\, d\Omega.$$

As $r_0 \to 0$, the partial derivatives $\partial h / \partial y^i$ tend to their values at p, and so

$$\frac{\partial F_3(2, t, p)}{\partial n} = \frac{v}{\omega_{N-1}} \frac{\partial h(p)}{\partial y^i} \int\limits_{\Omega_{N-1}} \lambda^i\, d\Omega = 0. \tag{10.4.28}$$

The vanishing of the normal derivative is a consequence of the evident relations $\int_\Omega \lambda^i\, d\Omega = 0$.

The solution formula may now be verified directly. For consider the function

$$u(P) = I^2 f(P) + V^2(P) + W^2(P), \tag{10.4.29}$$

where the single and double layers have the respective densities $g(p)$ and $h(p)$. From (10.4.12), (10.4.17), and (10.4.25) we have

$$\Delta u(P) = f(P) \quad (P \text{ in } D^P_{S}).$$

From (10.4.13), (10.4.20), and (10.4.28), we have

$$\frac{\partial u(p)}{\partial n} = g(p) \quad (p \text{ on } S^P),$$

and from (10.4.13), (10.4.19), and (10.4.27), we have

$$u(p) = h(p) \quad (p \text{ on } S^P).$$

Finally, we observe that if $N \geqslant 4$ is even the solution represents a wave travelling with clean-cut propagation. For N odd a residual wave is present.

10.5. Lorentz spaces of even dimension. According to this solution of the wave equation $\Delta u = f$, the propagation of waves is clean-cut in a Lorentz space of an even number N of dimensions. This holds, for example, when $N = 4$, the case of greatest physical interest. The propagation of light or sound in three spatial dimensions has this property, as indeed everyday experience demands.

We shall here explore some additional properties of the fractional hyperbolic potentials when N is even, and will show that the solution of the Cauchy problem can be put in a form which makes quite apparent the absence of any residual wave. If $u(P)$ is the solution which satisfies the conditions of Cauchy's problem, then according to (10.3.5) we have

$$I^\alpha u(P) = I^{\alpha+2}\Delta u(P) + V^{\alpha+2}(P) + W^{\alpha+2}(P). \tag{10.5.1}$$

In view of $\Delta I^{\alpha+2}u = I^\alpha u$, and the corresponding property of V^α and W^α, we have

$$I^\alpha u(P) = \Delta^{\frac{1}{2}N-1}[I^{\alpha+N}\Delta u(P) + V^{\alpha+N}(P) + W^{\alpha+N}(P)]$$

$$= \frac{\Delta^{\frac{1}{2}N-1}}{H_N(\alpha+N)}\left[\int\limits_{D_S^P} \Delta u \, r^\alpha \, dV + \int\limits_{S^P} \left(\frac{\partial u}{\partial n} r^\alpha - u \frac{\partial r^\alpha}{\partial n}\right) dS \right]. \tag{10.5.2}$$

This step is possible only if N is even, since only then is the exponent $\frac{1}{2}N - 1$ of Δ an integer. Let us regain the solution function $u(P)$ by passing to the limit $\alpha \to 0$ in this formula. We see at once that the limit of the first two terms is found without difficulty by replacing r^α by its limit $r^0 = 1$. However, in the third term the value obtained by setting $\alpha = 0$ in the differentiated expression $\alpha r^{\alpha-1}(\partial r/\partial n)$ is not fully determinate. The reason is that near the rim s^P of the initial surface, r approaches zero and the factor $r^{\alpha-1}$ becomes infinite as $\alpha \to 0$. We might expect from this that the limit will depend only on that part of the surface where r vanishes—that is, on s^P.

To calculate the limit of the double layer term, let us use the expressions (10.4.24) and (10.4.22). In (10.4.24) we must take the limit

$$\beta = \tfrac{1}{2}(\alpha - N) \to 0.$$

Thus we require to find

$$W^N(P) = \frac{2^{2-N}\sqrt{\pi}}{\Gamma(\tfrac{1}{2}N - \tfrac{1}{2})} \left[\frac{1}{\Gamma(\beta)} \int_0^1 F_3(N, t, P) t^{\frac{1}{2}(N-3)} (1-t)^{\beta-1} \, dt \right]_{\beta=0}. \quad (10.5.3)$$

According to (10.1.7) this limit is

$$\frac{2^{2-N}\sqrt{\pi}}{\Gamma(\tfrac{1}{2}N - \tfrac{1}{2})} F_3(N, 1, P) = \frac{2^{2-N}\sqrt{\pi}}{\Gamma(\tfrac{1}{2}N - \tfrac{1}{2})} \frac{1}{\Gamma(\tfrac{1}{2}N)\omega_{N-1}} \int_{\Omega_{N-1}} [u(q)]_{t=1} r_0^{N-2} \, d\Omega$$

$$= \frac{1}{\Gamma(N-1)\omega_{N-1}} \int_{\Omega_{N-1}} [u(q)]_{t=1} r_0^{N-2} \, d\Omega. \quad (10.5.4)$$

Here we have used (10.4.22), and the duplication formula for the gamma function. Now the values of $u(q)$ appearing in (10.5.4) are values on the rim s_P where $t = 1$; and the differential $r_0^{N-2} \, d\Omega$ is equal to the element of volume ds^P on the rim. We therefore conclude that

$$W^N(P) = \frac{1}{\Gamma(N-1)\omega_{N-1}} \int_{s^P} u \, ds. \quad (10.5.5)$$

This formula is also valid for N odd; it is not then of any use in the solution since the difference $N-2$ is also odd.

Returning to (10.5.2), we may now take the limit as $\alpha \to 0$. Noting that $H_N(N) = \Gamma(N-1)\omega_{N-1}$, we find the Riesz formula

$$u(P) = \frac{\Delta^{\frac{1}{2}N-1}}{\Gamma(N-1)\omega_{N-1}} \left[\int_{D_S^P} \Delta u \, dV + \int_{S^P} \frac{\partial u}{\partial n} \, dS + \int_{s^P} u \, ds \right]. \quad (10.5.6)$$

This concise expression shows directly the clean-cut nature of the wave propagation. Indeed, we need only observe that, since the integrands of the three terms no longer contain $r = r(P, Q)$, their only dependence on P is through the domains of integration. Since the operator $\Delta^{\frac{1}{2}N-1}$ is to be applied to these terms, it is evident that only the variation at the limits of these domains will contribute to the solution. Thus values of Δu in the interior of D_S^P, or values of $\partial u/\partial n$ in the interior of S^P, will not affect the value of $u(P)$, so that the domain of dependence of $u(P)$ is confined to the retrograde null cone C^P.

Evidently the case $N = 2$ must be considered separately. The rim s_P in this case consists of the two points p_1, p_2 where the characteristic curves through P intersect the initial curve S. The analogue of (10.4.22) for $N = 2$ is

$$F_3(\alpha, t, P) = \frac{1}{2\Gamma(\frac{1}{2}\alpha)}[h(p_1)r_0(p_1)^{\alpha-2}+h(p_2)r_0(p_2)^{\alpha-2}],$$

and it follows that

$$W^2(P) = \tfrac{1}{2}[h(p_1)+h(p_2)] = \tfrac{1}{2}[u(p_1)+u(p_2)].$$

The formula (10.5.6) now takes the form

$$u(P) = \tfrac{1}{2}\left[\int\limits_{D_S^P} \Delta u \, dV + \int\limits_{S^P} \frac{\partial u}{\partial n} \, dS + u(p_1)+u(p_2)\right], \qquad (10.5.7)$$

in which no differentiations appear. Thus the first two terms (though not their derivatives) depend upon the values of Δu and $\partial u/\partial n$ in the interior of the domain of dependence, and only the double layer contribution has the clean-cut wave property. In § 1.2 we derived this formula by elementary methods.

Evidently the clean-cut wave property is present in any of these solutions only as a consequence of the 'local' property of the one-dimensional Riemann–Liouville integral. This in turn is present only for $\alpha = 0$ or $\alpha = -n$, values which can be reached only by analytic continuation, except for $\alpha = 0$ which lies on the boundary of the region of convergence of the Riemann–Liouville integral and can be reached by taking a limit. Correspondingly, the fractional hyperbolic potentials exhibit the local property only for values of α at which the original integrals fail to converge. For the double layer $W^\alpha(P)$ the highest such value is $\alpha = N$, and we have calculated $W^N(P)$ in (10.5.5). For the single layer or the volume potential, the integrals as first given converge for $R(\alpha) > N-2$. We shall calculate $I^{N-2}f(P)$ as a limit.

According to (10.4.10), we have

$$I^{N-2}f(P) = \lim_{\alpha \to N-2} I^\alpha f(P)$$

$$= \frac{2^{-N+2}}{\Gamma(\frac{1}{2}N-\frac{1}{2})} \sqrt{\pi} \lim_{\substack{\alpha \to N-2 \\ \beta \to 0}} \int_0^1 F_1(\alpha, t, P)t^{\frac{1}{2}(N-3)}(1-t)^{\beta-1} \, dt$$

$$= \frac{2^{-N+2}}{\Gamma(\frac{1}{2}N-\frac{1}{2})} \sqrt{\pi} F_1(N-2, 1, P).$$

When $N = 4$, this formula yields the volume potential $I^2f(P)$ in the form (see (10.4.9))

$$I^2f(P) = \frac{1}{4\pi} \int_0^1 \sigma \, d\sigma \int_\Omega [f(Q)]_{t=1} r_0^2 \, d\Omega. \qquad (10.5.8)$$

The values of $f(Q)$ in the integrand are those taken on the cone C_S^P, where $r(P, Q) = 0$, and so

$$x^0 - \xi^0 = \sigma r_0 = \Big(\sum_{i=1}^{N-1} (x^i - \xi^i)^2 \Big)^{\frac{1}{2}} = \rho.$$

On the plane $x^0 = 0$, let the projection of C_S^P be denoted by $\bar{C}_S^P$, and let the area element be dA. Taking ρ, defined above, as a radial variable, θ^1 and θ^2 the usual angular variables, we have

$$dA = \rho^2 \, d\rho d\Omega,$$

and since $\rho = \sigma r_0$, $\quad \dfrac{dA}{\rho} = \rho \, d\rho d\Omega = r_0^2 \sigma \, d\sigma d\Omega.$

Substituting in (10.5.8), and writing $f(Q) = f(\xi^0, ..., \xi^3) = f(x^0 - \rho, \xi^1, \xi^2, \xi^3)$, we have

$$I^2f(P) = \frac{1}{4\pi} \int_{\bar{C}_S^P} f(x^0 - \rho, \xi^1, \xi^2, \xi^3) \frac{dA_0}{\rho}. \qquad (10.5.9)$$

This is the retarded potential of the classical theory of wave propagation.

10.6. Lorentz spaces of odd dimension. A formula analogous to (10.5.6), but in which the presence of a residual wave is evident, may be found when N is odd. For then the difference between $N+1$ and 2 is even, so that

$$u(P) = I^2\Delta u(P) + V^2(P) + W^2(P)$$
$$= \Delta^{\frac{1}{2}(N-1)}[I^{N+1}\Delta u(P) + V^{N+1}(P) + W^{N+1}(P)]. \qquad (10.6.1)$$

The integrals all converge for $\alpha = N+1$, and we may write the solution in the form

$$u(P) = \frac{\Delta^{\frac{1}{2}(N-1)}}{H_N(N+1)} \Big[\int_{D_S^P} r \Delta u \, dV + \int_{S^P} \Big(r \frac{\partial u}{\partial n} - u \frac{\partial r}{\partial n} \Big) dS \Big]. \qquad (10.6.2)$$

The integrands contain P through $r = r(P, Q)$ depending on P, and the indicated differentiations do not remove this dependence. Therefore the contributions to the solution from the interiors of D_S^P and of S^P constitute a residual wave.

This solution formula can be improved in the direction of reducing the number of differentiations to be performed, if we calculate the potentials for $\alpha = N-1$. The volume potential $I^{N-1}\Delta u(P)$ and the single layer $V^{N-1}(P)$ are represented by convergent integrals as above, but the integral for the double layer $W(P)$ diverges when $\alpha = N$. Let us work with the integral representation (10.4.24), with $\beta = \frac{1}{2}(\alpha - N)$. When $\alpha = N-1$, this integral diverges like $\int_0^1 (1-t)^{-\frac{3}{2}}\, dt$, unless the factor $F_3(\alpha, t, P)$ vanishes for $t = 1$. Let us add and subtract $F_3(\alpha, 1, P)$ in the integrand of $W^\alpha(P)$. We find

$$W^\alpha(P) = \frac{\sqrt{\pi}}{\Gamma(\frac{1}{2}N-\frac{1}{2})}\, \frac{2^{2-\alpha}}{\Gamma(\frac{1}{2}\alpha-\frac{1}{2}N)} \times$$

$$\times \int_0^1 [F_3(\alpha, t, P) - F_3(\alpha, 1, P)] t^{\frac{1}{2}(N-3)}(1-t)^{\frac{1}{2}(\alpha-N-2)}\, dt + \overline{W}^\alpha(P), \quad (10.6.3)$$

where $\overline{W}^\alpha(P)$ is the double layer of density $\bar{u}(q) = [u(t, \theta^\alpha)]_{t=1}$; the values taken by $u(q)$ on the rim s_P. We see from (10.4.22) that this term compensates exactly for the subtraction of $F_3(\alpha, 1, P)$ in (10.6.3). Supposing that $F_3(\alpha, t, P)$ is differentiable with respect to t, we see now that the integrand of (10.6.3) behaves like

$$[F_3(\alpha, t, P) - F_3(\alpha, 1, P)](1-t)^{\frac{1}{2}(\alpha-N-2)} = O\big((1-t)^{\frac{1}{2}(\alpha-N)}\big),$$

as $t \to 1$. Thus, as $\alpha \to N-1$, the integrand tends to infinity like $(1-t)^{-\frac{1}{2}}$, and the integral converges. Thus

$$W^{N-1}(P) = \frac{-1}{H_N(N-1)} \int_{S^P} (u - \bar{u})\, \frac{\partial}{\partial n}\left(\frac{1}{r}\right) dS + \overline{W}^\alpha(P), \quad (10.6.4)$$

the integral over S^P being convergent.

The term $\overline{W}^\alpha(P)$ may be calculated as follows. We have from (10.4.22)

$$F_3(\alpha, 1, P) = \frac{1}{\Gamma(\frac{1}{2}\alpha)\omega_{N-1}} \int_{\Omega_{N-1}} \bar{u}(q) r_0^{\alpha-2}\, d\Omega$$

$$= \frac{1}{\Gamma(\frac{1}{2}\alpha)\omega_{N-1}} \int_{s_P} u(q) r_0^{\alpha-N}\, ds, \quad (10.6.5)$$

since on s_P we have $r_0^{N-2}\, d\Omega = ds$. Since $F_3(\alpha, 1, P)$ is independent of t, we have from (10.4.24)

$$\overline{W}^\alpha(P) = \frac{2^{2-\alpha}}{\Gamma(\tfrac{1}{2}N-\tfrac{1}{2})} \sqrt{\pi} \left[\frac{1}{\Gamma(\beta)} \int_0^1 F_3(\alpha,1,P)t^{\tfrac{1}{2}(N-3)}(1-v^2)^{\beta-1}\,dt \right]_{\beta=\tfrac{1}{2}(\alpha-N)}$$

$$= \frac{2^{2-\alpha}}{\Gamma(\tfrac{1}{2}N-\tfrac{1}{2})} \sqrt{\pi} F_3(\alpha,1,P) \frac{B(\tfrac{1}{2}N-\tfrac{1}{2},\tfrac{1}{2}\alpha-\tfrac{1}{2}N)}{\Gamma(\tfrac{1}{2}\alpha-\tfrac{1}{2}N)}$$

$$= \frac{2^{2-\alpha}\sqrt{\pi}}{\Gamma(\tfrac{1}{2}\alpha-\tfrac{1}{2})} \frac{1}{\Gamma(\tfrac{1}{2}\alpha)\omega_{N-1}} \int_{s^P} u(q)r_0^{\alpha-N}\,ds. \tag{10.6.6}$$

In view of the duplication formula the numerical factor becomes the reciprocal of $\Gamma(\alpha)\omega_{N-1}$. In the integral we may replace r_0 by ρ the spatial distance, since $r_0 = \rho$ on s^P. Thus

$$\overline{W}^{N-1}(P) = \frac{1}{\Gamma(N-1)\omega_{N-1}} \int_{s_P} u(q)\frac{ds}{\rho} \tag{10.6.7}$$

The solution of the initial value problem may therefore be written

$$u(P) = \Delta^{\tfrac{1}{2}(N-3)}[I^{N-1}\Delta u(P) + V^{N-1}(P) + W^{N-1}(P)]$$

$$= \frac{\Delta^{\tfrac{1}{2}(N-3)}}{H_N(N-1)} \left[\int_{D_S^P} \Delta u \frac{1}{r}\,dV + \int_{S^P} \left(\frac{\partial u}{\partial n}\frac{1}{r} - (u-\bar u)\frac{\partial}{\partial n}\left(\frac{1}{r}\right) \right) dS \right] +$$

$$+ \frac{\Delta^{\tfrac{1}{2}(N-3)}}{\Gamma(N-1)\omega_{N-1}} \int_{s^P} u\frac{ds}{\rho}. \tag{10.6.8}$$

When $N = 3$, there are no differentiations to be performed, and we have the solution of the equation of cylindrical waves. Since

$$H_3(2) = \Gamma(2)\omega_2 = 2\pi,$$

we find the solution formula

$$u(P) = \frac{1}{2\pi}\left[\int_{D_S^P} \Delta u\frac{1}{r}\,dV + \int_{S^P} \left(\frac{\partial u}{\partial n}\frac{1}{r} - (u-\bar u)\frac{\partial}{\partial n}\left(\frac{1}{r}\right) \right) dS + \int_{s^P} u\frac{ds}{\rho} \right].$$

$$\tag{10.6.9}$$

If spatial polar coordinates (ρ,θ) are used, we have

$$u(P) = u(\rho,\theta,t), \qquad r^2 = t^2 - \rho^2,$$

and $\bar ds = \rho\,d\varphi$ is the element of length on the curve s_P.

10.7. The wave equation in a Riemann space. We have studied in detail the wave equation

$$\Delta u = \frac{1}{\sqrt{a}}\frac{\partial}{\partial x^i}\left(\sqrt{a}\,a^{ik}\frac{\partial u}{\partial x^k} \right) = 0, \tag{10.7.1}$$

in a Lorentz space, where the coefficients may be taken as constant, because the technique is simpler and also because the phenomenon of clean-cut wave propagation does not, in general, occur in curved spaces. We shall now describe how the fractional hyperbolic potential $I^\alpha f(P)$ may be constructed in a curved space of the normal hyperbolic signature. For simplicity, let us assume that all coefficients in (10.7.1) are analytic.

The construction hinges upon the form of the kernel $V^\alpha(P, Q)$, which, in our previous work, was

$$V^\alpha(P, Q) = \frac{r^{\alpha-N}(P, Q)}{H_N(\alpha)} = \frac{s^{\alpha-N}(P, Q)}{H_N(\alpha)},$$

the distance $r = r(P, Q)$ being the geodesic distance $s = s(P, Q)$. An important property of the kernel is, that operation upon it with Δ should reduce the index α by 2: $\Delta V^{\alpha+2}(P, Q) = V^\alpha(P, Q)$. We shall make use of this relation to find a suitable expression for $V^\alpha(P, Q)$ in this more general case. Following Riesz, let us assume that $V^\alpha(P, Q)$ can be written as a series of powers of the geodesic distance, with undetermined coefficients:

$$V^\alpha(P, Q) = \sum_{k=0}^{\infty} \frac{V_k(P, Q)s^{\alpha-N+2k}}{K_N(\alpha)L_N(\alpha+2k)}. \tag{10.7.2}$$

The coefficients $V_k(P, Q)$ will be determined by requiring

$$\Delta_Q V^{\alpha+2}(P, Q) = V^\alpha(P, Q).$$

We have from (4.5.23)

$$\Delta[s^{\alpha-N+2k+2}V_k]$$
$$= \frac{\alpha-N+2k+2}{2} s^{\alpha-N+2k}\left[2(\alpha-N+2k+\tfrac{1}{2}\Delta\Gamma)V_k+4s\frac{\partial V_k}{\partial s}\right]+s^{\alpha-N+2k+2}\Delta V_k. \tag{10.7.3}$$

From (4.5.31) we then find

$$\Delta[s^{\alpha-N+2k+2}V_k]$$
$$= (\alpha-N+2k+2)s^{\alpha-N+2k}\left[\left(\alpha+2k+s\frac{\partial\ln\sqrt{a}}{\partial s}\right)V_k+2s\frac{\partial V_k}{\partial s}\right]+s^{\alpha-N+2k+2}\Delta V_k$$

Thus

$$\Delta_Q V^{\alpha+2}(P, Q) = \sum_{k=0}^{\infty} \frac{\Delta(s^{\alpha-N+2k+2}V_k)}{K_N(\alpha+2)L_N(\alpha+2k+2)}$$

$$= \sum_{k=0}^{\infty} \frac{s^{\alpha-N+2k}(\alpha-N+2k+2)[\{\alpha+2k+s(\partial\ln\sqrt{a}/\partial s)\}V_k+2s(\partial V_k/\partial s)]}{K_N(\alpha+2)L_N(\alpha+2k+2)} +$$

$$+ \sum_{k=0}^{\infty} \frac{s^{\alpha-N+2k+2}\Delta V_k}{K_N(\alpha+2)L_N(\alpha+2k+2)}.$$

Replacing k by $k-1$ in the second sum, we find

$$\Delta V^{\alpha+2}(P, Q) = \sum_{k=0}^{\infty} \frac{s^{\alpha-N+2k}}{K_N(\alpha+2)} \left[\frac{\alpha+2+2k-N}{L_N(\alpha+2k+2)} \times \right.$$

$$\left. \times \left\{ \left(\alpha+2k+s \frac{\partial \ln \sqrt{a}}{\partial s}\right) V_k + 2s \frac{\partial V_k}{\partial s} \right\} + \frac{\Delta V_{k-1}}{L_N(\alpha+2k)} \right].$$

Let us require the factor $L_N(\alpha)$ to satisfy the condition

$$\frac{\alpha-N+2}{L_N(\alpha+2)} = \frac{1}{L_N(\alpha)}, \qquad (10.7.4)$$

then

$$\Delta V^{\alpha+2}(P, Q)$$

$$= \sum_{k=0}^{\infty} \frac{s^{\alpha-N+2k}}{K_N(\alpha+2)L_N(\alpha+2k)} \left[\left(\alpha+2k+s \frac{\partial \ln \sqrt{a}}{\partial s}\right) V_k + 2s \frac{\partial V_k}{\partial s} + \Delta V_{k-1} \right]. \qquad (10.7.5)$$

We may now choose the coefficients V_k, making them satisfy the recurrent relation

$$2s \frac{\partial V_k}{\partial s} + \left(2k+s \frac{\partial \ln \sqrt{a}}{\partial s}\right) V_k + \Delta V_{k-1} = 0. \qquad (10.7.6)$$

The square bracket in (10.7.5) then reduces to αV_k. We shall require the as yet undetermined factor $K_N(\alpha)$ to satisfy

$$\frac{\alpha}{K_N(\alpha+2)} = \frac{1}{K_N(\alpha)}. \qquad (10.7.7)$$

Then (10.7.5) becomes

$$\Delta_Q V^{\alpha+2}(P, Q) = \sum_{k=0}^{\infty} \frac{s^{\alpha+2k-N}V_k}{K_N(\alpha)L_N(\alpha+2k)} = V^{\alpha}(P, Q), \qquad (10.7.8)$$

and the required condition is satisfied.

The recursion formula (10.7.6) has the form of an ordinary differential equation of the first order, the independent variable $s = s(P, Q)$ varying with Q. For $Q = P$, $s = 0$, and the V_k must all have finite values. For $V_0(Q, Q)$ we choose the value unity (as in the Lorentzian case), with $V_{-1} \equiv 0$. It then follows that

$$V_0(P, Q) = \left(\frac{a(P)}{a(Q)}\right)^{\frac{1}{4}}, \qquad (10.7.9)$$

and, generally, for $k \geqslant 1$,

$$V_k(P, Q) = -V_0 s^{-k} \int_0^s \frac{\sigma^{k-1}}{2V_0(\sigma)} \Delta V_{k-1}(\sigma)\, d\sigma. \qquad (10.7.10)$$

Thus the coefficients are altogether determined by (10.7.6) and the condition of finiteness for $s = 0$.

The equations (10.7.4) and (10.7.7) are satisfied if

$$K_N(\alpha) = K\, 2^{\frac{1}{2}\alpha}\Gamma(\tfrac{1}{2}\alpha); \qquad L_N(\alpha) = L\, 2^{\frac{1}{2}\alpha}\Gamma(\tfrac{1}{2}\alpha - \tfrac{1}{2}N + 1).$$

Writing $\qquad H_N(\alpha, k) = K_N(\alpha)L_N(\alpha + 2k),$

and requiring that $H_N(\alpha, 0) = H_N(\alpha)$, which must be true in the Lorentzian case, we find

$$H_N(\alpha, k) = \pi^{\frac{1}{2}(N-2)}2^{\alpha+k-1}\Gamma(\tfrac{1}{2}\alpha)\Gamma(\tfrac{1}{2}\alpha - \tfrac{1}{2}N + k + 1). \qquad (10.7.11)$$

The convergence of the series for $V^\alpha(P, Q)$ can be demonstrated, for s sufficiently small, by the method of dominant power series. As the details are lengthy, we shall not include any such proof here.

The fractional potential is again defined as

$$I^\alpha f(P) = \int_{D_s^P} V^\alpha(P, Q)f(Q)\, dV_Q, \qquad (10.7.12)$$

and, as in § 10.3, similar definitions hold for the potentials of the surface layers $V^\alpha(P)$ and $W^\alpha(P)$. The possibility of analytic continuation to values of α for which the integral (10.7.12) diverges can be demonstrated just as before. One difference may be noted, however. Only the first few leading terms in the series (10.7.2) will become infinite when $s \to 0$; and all terms but these lead to convergent terms in $I^\alpha f(P)$. Thus these convergent integrals depend explicitly on values of $f(Q)$ in the interior of D_s^P and contribute to a residual wave in the solution of the initial value problem.

The identity relation $I^0 f(P) = f(P)$ is also valid. In fact only the first term in the series for $V^\alpha(P, Q)$ contributes to $I^0 f(P)$, the convergent higher terms dropping out because of the factor $\Gamma(\tfrac{1}{2}\alpha)$ in the denominator, and the other divergent terms containing a higher power of s which is zero at the apex P of the cone. The differentiation formula

$$\Delta I^{\alpha+2}f(P) = I^\alpha f(P)$$

is again valid, as a consequence of the corresponding relation

$$\Delta_P V^{\alpha+2}(P, Q) = V^\alpha(P, Q).$$

This latter holds because $V^\alpha(P, Q)$ is symmetric in its two argument points P and Q. The proof of symmetry is rather long, and will be omitted. We also mention without proof that the rule of composition

$$I^\alpha I^\beta f(P) = I^{\alpha+\beta}f(P)$$

holds, as a consequence of the formula

$$\int_{D_Q^P} V^\alpha(P, R)V^\beta(Q, R)\, dV_R = V^{\alpha+\beta}(P, Q),$$

which is the present analogue of (10.2.12).

The generalized potential (10.7.12) is used in the solution of the initial value problem, just as in § 10.3. Applying Green's formula, and performing the analytic continuation to $\alpha = 0$, we again find that the solution formula is given by (10.3.9). Verification of the solution follows § 10.4, since it is easy to show that the higher terms in the series (10.7.2) for $V^\alpha(P, Q)$ do not contribute to those values of the potentials which are concerned.

This concludes our brief description of the wave equation in a curvilinear Riemann space. For additional details of the Riesz technique, and of other integration methods, the reader is referred to the bibliography.

Exercise 1. The operator I^α is symbolically equal to $\Delta^{-\frac{1}{2}\alpha}$. Show that the operator I_k^α corresponding to the equation $\Delta u + k^2 u = 0$ is equivalent to

$$\sum_{n=0}^{\infty} \binom{-\frac{1}{2}\alpha}{n} k^{2n} I^{\alpha+2n}.$$

If I^α has kernel $r^{\alpha-N}/H_N(\alpha)$, show that I_k^α has the kernel

$$\frac{1}{\pi^{\frac{1}{2}(N-2)} 2^{\frac{1}{2}(\alpha+N-2)} \Gamma(\frac{1}{2}\alpha)} \left(\frac{r}{k}\right)^{\frac{1}{2}(\alpha-N)} J_{\frac{1}{2}(\alpha-N)}(kr).$$

Exercise 2. Show that the solution of $u_{xy} + k^2 u = 0$ which is equal to $\psi(x)$ when $y = 0$ and to $\varphi(y)$ when $x = 0$ is given by Laplace's formula

$$u(x, y) = \int_0^y J_0[k\sqrt{\{x(y-s)\}}]\varphi'(s)\, ds + \int_0^x J_0[k\sqrt{\{y(x-s)\}}]\psi'(s)\, ds.$$

Find a similar formula for the non-homogeneous differential equation with $f(x, y)$ on the right-hand side. Assume $\varphi(0) = \psi(0) = 0$.

Exercise 3. Show that the solution of $u_{xx} + u_{yy} - (1/c^2)u_{tt} = 0$ which satisfies $u(x, y, 0) = g(x, y)$, $u_t(x, y, 0) = G(x, y)$, is given by

$$2\pi c u(x, y, t) = \frac{\partial}{\partial t} \int_0^{ct} \int_0^{2\pi} \frac{g(\xi, \eta) r\, dr d\theta}{\sqrt{(c^2 t^2 - r^2)}} + \int_0^{ct} \int_0^{2\pi} \frac{G(\xi, \eta) r\, dr d\theta}{\sqrt{(c^2 t^2 - r^2)}},$$

where $x - \xi = r\cos\theta$, $y - \eta = r\sin\theta$. Show that this solution agrees with (10.6.9).

Exercise 4. Show that the equation

$$\frac{1}{c^2} u_{tt} = u_{rr} + \frac{1}{r} u_r$$

has the solution

$$u = \frac{1}{2\pi} \int\limits_0^\infty f\left(t - \frac{r}{c}\cosh u\right) du = \frac{c}{2\pi} \int\limits_{-\infty}^{t-(r/c)} \frac{f(\tau)\,d\tau}{[c^2(t-\tau)^2 - r^2]^{\frac{1}{2}}},$$

where $f(t)$ is an arbitrary function. Show that

$$\lim_{r \to 0}\left(-2\pi r\,\frac{\partial u}{\partial r}\right) = f(t).$$

Deduce the form of waves on a plane sheet of water caused by a disturbance of strength $f(t)$ at the origin.

Exercise 5. Show that all solutions of $u_{xx} + u_{yy} + u_{zz} = (1/c^2)u_{tt}$, which depend on r and t only, may be represented in the form

$$u = \frac{1}{r}[f(r+ct) + g(r-ct)].$$

Exercise 6. Let $g(x, y, z)$ be a twice differentiable function, and let the mean value of $g(x, y, z)$ on the surface of a sphere of radius r be denoted by $M_r(g)$:

$$M_r(g) = \frac{1}{4\pi} \int\limits_0^{2\pi}\int\limits_0^\pi g(x + r\cos\theta\cos\varphi,\ y + r\cos\theta\sin\varphi,\ z + r\sin\theta)\sin\theta\,d\theta d\varphi.$$

Show that $tM_{ct}(g)$ is a solution of $u_{xx} + u_{yy} + u_{zz} = (1/c^2)u_{tt}$.

Exercise 7. Show that the solution of $u_{xx} + u_{yy} + u_{zz} = (1/c^2)u_{tt}$, with $u(x, y, z, 0) = g(x, y, z)$; $u_t(x, y, z, 0) = h(x, y, z)$ is

$$u = tM_{ct}(h) + \frac{\partial}{\partial t}[tM_{ct}(g)].$$

BIBLIOGRAPHY

1. B. B. Baker and E. T. Copson, *The mathematical theory of Huygens' principle* (Oxford, 1939).
2. H. Bateman, *Partial differential equations* (Cambridge, 1932).
3. S. Bergman and M. Schiffer, *The kernel function in mathematical physics* (New York, 1953).
4. D. L. Bernstein, *Existence theorems in partial differential equations* (Princeton, 1950).
5. L. Bieberbach, *Theorie der Differentialgleichungen* (Berlin, 1930).
6. C. Carathéodory, *Variationsrechnung und partielle Differentialgleichungen erster Ordnung* (Berlin, 1935).
7. E. Cartan, *Leçons sur les invariants intégraux* (Paris, 1922).
8. R. Courant, *Dirichlet's principle* (New York, 1950).
9. R. Courant and D. Hilbert, *Methoden der mathematischen Physik*, 2 vols. (Berlin, 1931, 1937).
10. L. P. Eisenhart, *Riemannian geometry* (Princeton, 1927).
11. P. Frank and R. von Mises, *Die Differentialgleichungen der Mechanik und Physik*, 2 vols. (Berlin, 1930, 1935).
12. E. Goursat, *Cours d'analyse mathématique*, 3 vols. (Paris, 1902–15).
12a. E. Goursat, *A course of mathematical analysis* (trans. Hedrick), vol. I, vol. II, parts I and II (Boston, 1917).
13. J. Hadamard, *Lectures on Cauchy's problem* (New Haven, 1923).
14. D. Hilbert, *Grundzüge einer allgemeinen Theorie der linearen Integralgleichungen* (Berlin, 1913).
15. E. Hille, *Functional analysis and semi-groups* (New York, 1948).
16. E. W. Hobson, *Theory of spherical and ellipsoidal harmonics* (Cambridge, 1931).
17. W. V. D. Hodge, *Theory and applications of harmonic integrals* (Cambridge, 1941).
18. E. L. Ince, *Ordinary differential equations* (New York, 1944).
19. E. Kamke, *Differentialgleichungen reeller Funktionen* (Leipzig, 1930).
20. O. D. Kellogg, *Foundations of potential theory* (Berlin, 1929).
21. C. Lanczos, *The variational principles of mechanics* (Toronto, 1949).
22. P. Lévy, *Problèmes concrets d'analyse fonctionnelle* (Paris, 1951).
23. J. Leray, *Hyperbolic differential equations* (Princeton, 1954).
24. W. V. Lovitt, *Linear integral equations* (New York, 1924).
25. M. Riesz, 'L'intégrale de Riemann–Liouville et le problème de Cauchy', *Acta Math.* 81 (1949) 1–223.
26. M. Schiffer and D. C. Spencer, *Functionals of finite Riemann surfaces* (Princeton, 1954).
27. W. Schmeidler, *Integralgleichungen mit Anwendungen in Physik und Technik* (Leipzig, 1950).
28. J. A. Schouten and W. v. der Kulk, *Pfaff's problem and its generalizations* (Oxford, 1949).
29. I. N. Sneddon, *Fourier transforms* (New York, 1949).
30. A. Sommerfeld, *Partial differential equations* (New York, 1949).
31. W. Sternberg and T. L. Smith, *Theory of potential and spherical harmonics* (Toronto, 1943).
32. J. A. Stratton, *Electromagnetic theory* (New York, 1941).

33. J. L. SYNGE and A. SCHILD, *Tensor calculus* (Toronto, 1949).
34. V. VOLTERRA and J. PERÈS, *Théorie générale des fonctionnelles* (Paris, 1936).
35. A. G. WEBSTER, *Partial differential equations of mathematical physics* (Boston, 1925).
36. H. WEYL, *Die Idee der Riemannschen Fläche* (Leipzig, 1913).
37. E. T. WHITTAKER and G. N. WATSON, *A course of modern analysis*, 4th ed. (Cambridge, 1927).

Chapter references:

Ch. I: (4), (9, vol. ii, ch. i), (12a, vol. ii, part II), (13, book I), (18, ch. ii), (33, ch. i).
Ch. II and Ch. III: (6, part II), (7), (9, vol. ii, ch. ii), (12a, vol. ii, part II), (21, ch. viii).
Ch. IV: (9, vol. i, ch. i, vol. ii, ch. iii), (10, ch. ii), (33, ch. ii).
Ch. V: (8), (14, ch. xviii), (20), (25), (36), (37, ch. xx).
Ch. VI: (14), (23), (26), (30), (34).
Ch. VII: (3), (14), (20).
Ch. VIII: (2), (9, vol. i, ch. iv), (15), (16), (29), (30).
Ch. IX: (1), (9, vol. ii, chs. v, vi), (13), (35).
Ch. X: (1), (9, vol. ii, chs. v, vi), (13), (23), (25).

INDEX

adjoint operator, 83.
  system, 40.
alternative, 133.
analytic continuation, 219, 225.
  equations, 9.
  functions, 8, 171, 217.
  solution, 113.
Ascoli, 166.
auxiliary condition, 7.

Bessel function, 87, 154.
  inequality, 145.
bicharacteristic, 200.
bilinear series, 144, 185.
boundary condition, 7.

Cauchy method, 48.
  problem, 224.
characteristic, 10, 72.
  cone, 204.
  curve, 26.
  direction, 25.
  displacement, 48.
  equation, 201.
  form, 73.
  manifold, 33, 54.
  ray, 201.
  root, 74, 138.
  strip, 49.
  surface, 195, 200.
  surface element, 43.
  value, 130.
Christoffel symbol, 90.
closed space, 115.
commutator, 39.
compact, 165.
complete integral, 55, 65,
  system, 37.
completely integrable, 17, 44 ,67.
complex variable, 217.
composition law, 221, 241.
configuration space, 21.
conjugate function, 172.
  tensor, 75.
constant coefficients, 85.
contravariant, 21.
  vector, 48.
coordinate, 20.
correctly set, 8, 100, 212.
covariant, 20.
curvature, 84, 118.
curve, 21.

dependent variable, 1.
determinant, 126.
diffusion, 191.
direct cone, 204.
Dirichlet integral, 84, 99.
  problem, 7, 120, 151, 170, 182.
dispersion, 212.
divergence, 78.
domain functional, 157, 159, 161, 173.
domain of dependence, 215.
dominant, 12.
double, 117.
  layers, 113, 225, 231.
dual system, 42.

eigenfunction, 100, 138.
eigenvalue, 130, 141, 186.
elliptic, 72.
envelope, 6, 58.
equivalent systems, 37.
Euclidean geometry, 35.
Euler equation, 90, 202.
extended system, 37.
exterior problem, 153.

first integral, 27, 50.
flat space, 85.
Fourier series, 144, 189.
fractional integral, 217.
Fredholm, 120, 125.
  alternative, 133.
  equation, 121, 125, 151.
function, of complex variable, 10, 11, 218.
functional determinant, 21, 32, 53.
  relation. 4.
fundamental singularity, 86.
  solution, 86, 104, 147, 177.
  tensor, 71.

Gauss formula, 81.
  theorem, 100.
general integral, 27, 30.
  solution, 9, 10.
geodesic, 88, 202.
geometrical space, 21.
global, 9.
gradient, 22, 75.
Green's formula, 80, 83, 84, 119, 224.
  function, 157, 173, 181, 187.

Hamiltonian, 203.
  system, 59.

Lightning Source UK Ltd.
Milton Keynes UK
UKHW010003210722
406167UK00001B/163